ATOMIC ENERGY: A NEW START

A 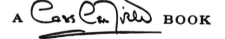 BOOK

*Books by David E. Lilienthal*

1/13/82

# David E. Lilienthal

# ATOMIC ENERGY: A NEW START

HARPER & ROW, PUBLISHERS, New York
Cambridge, Hagerstown, Philadelphia, San Francisco
London, Mexico City, São Paulo, Sydney

1817

FIRST EDITION

Designer: Sidney Feinberg

Library of Congress Cataloging in Publication Data

Lilienthal, David Eli, 1899–
Atomic energy, a new start.
Includes index.
1. Atomic power-plants—United States.
2. Atomic power—United States. 3. Atomic energy industries—United States. 4. Atomic
energy policy—United States. I. Title. TK1343.L54     621.48'3'0973      79-3668
ISBN 0-06-012617-5

80 81 82 83 10 9 8 7 6 5 4 3 2 1

This book is dedicated to
ALBERT EINSTEIN
whose awesome scientific discoveries have
shaken the foundations of our world.

# CONTENTS

# ACKNOWLEDGMENTS

Once again, after a publishing association of more than thirty-five years, I am proud to acknowledge my debt to Cass Canfield for proposing the writing of this account of my life with the atom, and for the way in which, in his person and his career, he has continued to exemplify the best standards of the embattled profession of publishing.

Through the many intense weeks of the thinking and writing of this book I have enjoyed the intellectual companionship, the forthright modern critical judgment, and the editorial assistance of my son David. I take full responsibility for the conclusions of the book, but for the spirit of an open mind which I hope the book represents, I acknowledge with satisfaction my debt to my son.

In the development of my ideas about atomic energy over more than three decades, I have benefited from many discussions with some of those imaginative and responsible people who have created this new technology and guided its course. Their names are too numerous to recite, but I wish to record here my indebtedness to them.

For those basic concepts of science and the public welfare by which I have lived and worked, I am deeply indebted to the greatest humanist and life scientist I have ever known, the late and beloved Harcourt A. Morgan, my colleague and teacher in the early days of the Tennessee Valley Authority.

I acknowledge with gratitude and admiration the skill and stamina of my longtime secretary and friend, Mildred Baron, who typed a succession of versions of this work and prepared the final manuscript for the printer.

I am indebted to Corona Machemer, my editor at Harper & Row, for her perceptive suggestions and criticism.

D.E.L.

# PREFACE

In the early years of this century, when I was growing up in the small Indiana town of Valparaiso, the major inventions and discoveries that have completely changed the world we live in were still in their infancy.

In Valparaiso there was only one automobile—the doctor's—which we boys would gather round and stare at in wonder. Horses were what we were used to; almost every family had one, and when farmers came to town on Saturdays, the courthouse square was lined with their teams and wagons. People had no electricity in their homes; they read by kerosene lamps. Whoever wanted a bath went outside to the well, pumped up some water, and heated it on the wood-burning kitchen range.

By and large, life in our town and in small towns across the whole of America proceeded at a slow, quiet pace. Horses and human muscle, not electricity and gasoline, were the chief sources of energy.

All that has changed.

Like others of my generation, I have seen the most revolutionary transformation in all human history of the way people live—a process that is rushing ahead pell-mell, at an ever increasing rate.

Much of this vast change centers on *energy,* most recently and most profoundly on the most controversial and fearsome of all forms of energy, the fundamental force of nature found in the nucleus of the atom—a source of energy about which there have arisen passionate opposition and equally intense support the world over.

At four o'clock on the morning of March 28, 1979, the alarm lights began to flash in the control room of Unit Two of the Metropolitan Edison atomic energy plant on Three Mile Island in the Susquehanna River, eleven miles southeast of Harrisburg, Pennsylvania.

This marked the beginning of the worst nuclear accident scare in history. Thousands of persons in the surrounding four-county area evacuated their homes. Thousands of others prepared to leave the area. For six days America and the entire world watched and waited with suspended breath as nuclear engineers gingerly executed a series of operations to cool down the plant's dangerously overheated nuclear core.

The reaction to the accident was immediate and intense. Nearly one hundred thousand protesters gathered in Washington in the largest political demonstration since the Vietnam war. There were powerful echoes abroad among the strong antinuclear forces in industrial countries like France, Germany, Italy, and Japan—and in Sweden, where the previous year a government had been forced to resign over the atomic energy issue. Thus Three Mile Island dramatized the emergence of nuclear power as the storm center of the biggest issue on the world scene, short of atomic war—energy.

For the ordinary American Three Mile Island raised some fundamental—and unanswered—questions about energy supply and atomic power.

Would there be other nuclear reactor accidents—worse ones? Was nuclear power simply too risky? Should we stop building new plants—and close down some or all of those we had?

It was some measure of the confusion and fear about atomic energy in this country that we found ourselves asking such questions nearly a quarter century after the beginning of the nuclear energy program—with nuclear reactors operating in no fewer than twenty-five states, making a crucial contribution to our electrical energy requirements.

For better or worse, I was one of the founders of the United States atomic policy. In 1946, soon after atomic bombs were dropped on Japan, when I was still chairman of the country's largest electricity producer, the Tennessee Valley Authority, I was called on by the President and the Secretary of State to head a special board of consultants to develop an American plan for international control of atomic energy and atomic weapons. Shortly thereafter, I was appointed by President Truman as the first chairman of the newly created civilian U.S. Atomic Energy Commission, and served until my resignation in early 1950. Since then, as a private citizen, to avoid any question of conflict of interest, I have refrained from entering into any tie or commitment to any of the public or private groups concerned with nuclear development. I have, however, spoken out from time to time, often critically, on atomic energy policies. An early advocate of peaceful atomic energy, I was also among the first to raise warnings about atomic waste disposal and other nuclear dangers, and in 1963, initiated the opposition to the proposed location of an atomic energy plant within the borough of Queens, New York City.

I have been engaged all my life in energy development: as a founding director of the TVA and its chairman in a period when it was synonymous with regional river control and hydroelectric power; as chairman of the AEC; and, after I left the Atomic Energy Commission, as founder of a private development company which became involved in energy projects in many parts of the world.

The time has come for me to look back over what I have seen and done, to sort out the pieces, and to set down some conclusions. I do this not in a reminiscent mood but rather in a search for whatever light my personal experience and working responsibilities throw on the baffling and unresolved energy problems that have come to dominate the lives and prospects of Americans and the other peoples of the industrially developed world.

To write these chapters, I have come to the cottage on the island of Martha's Vineyard where my wife and I have spent our summers for the past thirty years, an isolated hilltop whose tranquillity reminds me of that little Indiana town of my boyhood. All I see on this summer day is serene. To the north spreads the unbroken green of scrub oak and pine woods with the dense blue of Vineyard Sound beyond it. To the south lies a distant azure line—the Atlantic. There is no sign of human habitation; there is no hint of discord or trouble in anything I see or hear on this quiet summer day. But of course I know that all this is a blessed illusion. Even in the Vineyard towns there are gasoline lines, worries about heating oil shortages for the winter, and the further, deeper anxieties about economic recession and whether the future will bring peace or war.

Much of our fear stems from the terrible consequences of the scientific discoveries of this atomic age—the literally tens of thousands of atomic weapons now poised, ready to destroy the world, and the scores of atomic energy plants that produce electricity for human use but pose a threat because of their radiation hazards. I have lived for many years with these hard realities. The present book is based on the conclusions I have drawn from firsthand experience and responsibilities.

My conviction is that it is neither sensible nor useful to reflect on atomic energy apart from a full consideration of energy from all sources, including those in their promising technical infancy, such as solar energy. Nor is it wise to arrive at short-term conclusions about atomic energy. The atomic discoveries of only a few years ago are having profound consequences on our lives today, and they will continue to do so for many years. The atom is forcing a reevaluation of Western industrial society; it is affecting the lives of people everywhere.

As I think of the next decade or two, it becomes increasingly clear to me that far deeper questions are involved than whether the atom is or can be made safe as a new source of electrical energy, or whether that form of energy is more economic than other forms. These questions are relevant, of course, and I shall discuss them. But the chief value of the reexamination I am undertaking may lie in whatever by-product of insight it can yield into the human values and political institutions, national and international, which, before the atomic discoveries, had come to be accepted as the firm and settled conditions of life.

I believe that atomic energy has done more than challenge these assumptions. It has unsettled all things mortal. It has enlarged and magnified man's understanding of his own capacity to comprehend the universe and how it is put together.

The true "crisis" that confronts the world today is not the frequently discussed "energy crisis," nor is it a crisis of political or business "confidence." The present crisis is that man's very fate has been challenged by the consequences of his own imagination and curiosity. This is the true significance of the atomic adventure, and it becomes all the more evident the more we are able to see with perspective some of the irreversible turnings on the road of history that we have followed since the first atomic discoveries—a process that is still going on, and from which there is no turning back.

Living with the atom is now a condition of life—or death—for all humanity for centuries to come.

D.E.L.

*Topside*
*Martha's Vineyard, Mass.*

*88 Battle Road*
*Princeton, N.J.*

ATOMIC ENERGY: A NEW START

# 1

# A HOPE THAT FOUNDERED

In the late afternoon of April 3, 1947, I was in the Oval Room of the White House. As head of the newly created civilian Atomic Energy Commission, I had come to report to President Truman on the atomic energy establishment we had just taken over from the Army's wartime Manhattan District. At this time it was assumed by everyone, the President included, that America had a supply of atomic bombs. In fact, Winston Churchill was declaiming that it was our atomic "stockpile" that restrained the Soviet Union from moving in on an otherwise defenseless Europe.

What we of the new AEC had just discovered, in taking our inventory of nuclear materials and devices inherited from the Army, was that this defense did not exist. There was no stockpile. There was not a single operable atomic bomb in the "vault" at the Los Alamos atomic arsenal. Nor could there be one for many months to come.

This news was top secret, the biggest secret of that time, so secret that I did not commit it to paper, even as a part of the AEC secret archive.

As I gave President Truman this shocking information, the impact of the news was evident in his expression. I wrote in my private journal that night (without hinting at the nature of the subject concerned), "He turned to me, a grim, gray look on his face, the lines from his nose to his mouth visibly deepened."

At the end of this sobering meeting we shook hands, as usual, and made our goodbyes. I noted in my journal entry: "The President was rather subdued and thoughtful-looking—his customary joke on parting was missing. Then, in that abrupt, jerky way of talking, he said, 'Come in to see me any time, just any time. I'll always be glad to see you. You have the most important thing there is. You must make a blessing of it or [and a half-grin as he pointed to a large globe in the corner of his office] we'll blow that all to smithereens.' "

President Truman's injunction to "make a blessing" of the atom rang in my ears from that day on. In that homely phrase he expressed the hope and purpose of all Americans, scientists and average citizens alike. This became our job, our mission: to find a way to convert the virtually unlimited power of nuclear fission into electric energy, for the benefit of all mankind.

A whole generation has gone by since the dawning of what was so hopefully called the atomic age. America has spent millions and millions of dollars on atomic development. Many of the most talented men and women in science, industry, the universities, and government have devoted their careers to this field. Nuclear plants have been designed, built, and placed in operation. In state after state and city after city Americans have come to depend heavily on nuclear energy.

Until lately we had reason to think that we had in fact realized the great ambition, to "make a blessing" of the atom.

We were wrong.

The technical method chosen for producing electricity from fission has proven to be far from an unmixed blessing. While it produces electricity, it is cursed with radiation hazards that have caused great unease in the public mind, and it has led to passionate antinuclear protest movements in this country and in virtually every industrial nation in the world.

There is a bitter irony in my reflections about what has happened since I told President Truman about that empty A-bomb vault at Los Alamos. Bombs we now have, by the thousands, as do the Russians. But we do not yet have a safe method of producing atomic electricity.

This is the hard fact that America must face.

My object in writing this book is to try to explain how the world arrived at this turning point in the history of atomic energy development.

More than that, I will suggest what we can do to move ahead on a better course, which some day may make a reality of that hoped for "blessing" of which Harry Truman spoke to me so fervently on that April afternoon in Washington, long years ago.

# 2

# DO WE NEED NUCLEAR ENERGY?

Some who oppose nuclear power assert that we can do without it. They maintain that we can satisfy our energy needs by cutting down on our consumption, by instituting more efficient energy uses in factories and homes, and by pushing hard to develop nonnuclear energy sources, especially coal and solar energy.

This position is advanced seriously, and I have examined it seriously; it is fundamental to the question of nuclear power.

*Oil* is far and away our major source of energy today. How unpredictable and bitterly controversial this source is! Only a few years ago our laws forbade the importation of foreign oil without a federal license, in order to "protect" our domestic oil supplies. Then we were paying the Middle East producers an unconscionably low $2 a barrel; at this writing, a barrel of oil costs nearly $30. Everyone knows how dependent we have become on foreign oil and how political and even military uncertainties are a constant threat to our oil supplies from the OPEC countries.

As crude oil prices skyrocket, the economic incentives and technical prospects brighten for getting more oil from "exhausted" American wells. And new finds outside the continental United States are reported almost daily. But it is clear that the era of plentiful, reliable, and cheap petroleum is over. America's basic energy policy is to phase out as soon as possible its continuing dependence on oil from the Middle East, a reliance that has become the economic jugular vein of the industrialized West. One need not be reminded of the economic shock wave of the Arab oil embargo of October 1973, the overnight multiplying of crude oil prices imposed by the OPEC suppliers, the complete cutoff by Iran of oil to America in 1979, and in 1980 the ominous threat to oil supplies—and to world peace—of the military drive of the Soviet Union toward the Persian Gulf. That a conflict over access to the oil of the Middle East may create a military confrontation—that is, war—between America and the Soviet Union is no longer a remote rhetorical possibility.

In 1978 *coal* was the source for roughly 50 percent of the electric power used by the nation as a whole. This percentage will not be materially different by 1988; for the present the country will rely heavily on coal for its electricity generation.

We are constantly reminding ourselves that America has immense deposits of coal that could be used as a source of energy. In the 1960s I was one of those who were active proponents of large-scale utilization of the great deposits of low-sulphur coal in our western mountain states. I also pressed for a program for obtaining oil from coal, using well-established technologies. That proposal for a synthetic oil program and similar proposals were stalemated. Years later, in 1979, the administration aggressively promoted synthetic fuel as a practical alternative to imported petroleum. Although there is still room for doubt about whether the technical and economic questions can be resolved, the prospects for success appear to be good.

While the knowledge that it has almost unlimited reserves of coal has given America a sense of security with respect to its long-term energy supply, not only for electricity but for liquid fuel, there is a serious flaw in this reliance on coal. The poisoning of the air by the burning of coal is not the only hazard; recent scientific evidence indicates that the massive combustion of coal —indeed, of any fossil fuel, such as petroleum—creates a potential hazard even greater than the impairment of the health of individuals. The very stability of the global environment, our climate, may be threatened by the inhibiting effect on cosmic radiation of the great volume of carbon dioxide and nitrogen oxide released by the worldwide combustion of fossil fuels. There may well be technical solutions or mitigating answers to such an awesome problem. But with the underdeveloped countries— China, for one—only beginning the large-scale burning of coal for energy and the combustion of petroleum for transportation, it may no longer be sensible for America to rely with such confidence on its possession of vast resources of coal. The dangers may be even more far-reaching than those of atomic energy plants.

*Hydroelectricity,* produced from the power of falling water impounded by dams on our rivers and streams, is a "clean" and renewable energy resource. Thousands of electric generating possibilities still exist in this country. Most are small; together they add up to a respectable total. When I first espoused the cause of small hydro in September 1977,* there were many, including President Carter, who gave it short shrift, asserting that we have gotten everything we can out of hydro power. The skeptics were wrong; even though small, "low-head" hydro has definite limits and cannot be counted on to supply more than a fraction of our needs—though a growing fraction—the potential overall is considerable for obtaining hydroelectricity from the still undeveloped rivers and streams of the states of the American Northeast. And of all the United States, this is the region that is most dependent on imported oil for its electricity supply.

Across the American border, in the Dominion of Canada and reaching into the Arctic, is an underutilized hydroelectric re-

---

*"Lost Megawatts Flow over Nation's Myriad Spillways," *Smithsonian,* September 1977, p. 82.

source of enormous magnitude. The pioneer in North America in the development of hydroelectric systems was a Canadian, the great Sir Adam Beck. Hydro of Ontario, his creation, is one of the world's leaders in the managerial and technical aspects of water power. On a recent brief visit to the backcountry of the Province of Quebec, I had it borne in on me—no novice in the large-scale utilization of river systems—that the first stage alone, now completed, of Quebec's giant James Bay hydro development represents one of the largest increments of water power anywhere in the world. James Bay is within feasible transmission distance of energy-hungry New England, and negotiations are well under way for the purchase of a portion of this electricity, to add to the Canadian water power already flowing into the American Northeast.

*Solar energy* is a new and exciting field that promises much for the future. Its broadest application thus far has been as an adjunct to heating energy; it has already established a tiny but growing place in our economy. Whether further developments will justify the sanguine predictions of its enthusiasts, except at fantastic cost, has seriously been questioned by a special committee of the nation's top independent scientific body, the National Academy of Sciences (January 1980). Ill-considered euphoria for solar energy as a solution to all our energy needs could be as damaging as the early euphoria for atomic energy was.

*Fusion* remains an unresolved possibility for the next century.

*Conservation and efficiency of energy use* offer, by all odds, the most immediate and practical one-shot means of increasing the available energy supply for the short term. Ever since the 1974 pioneer Ford Foundation energy report by S. David Freeman,* now chairman of TVA, there have been striking calculations and

---

*See his classic study, *Energy: The New Era*, New York: Random House, 1974.

heartening cases of energy savings, notably in home and water heating and the insulation and redesign of structures. And certainly the group use of automobiles and improved mass transit will save on gasoline, as will the use of smaller, lighter cars. Persistent and hopeful as we must be about conservation and the avoidance of waste, we must realistically take into account the established and cherished habits of energy use of the American people and the motorized dynamics of our entire economy. The fact is that major, across-the-board savings through conservation thus far have been largely only talked about; conservation does not offer an easy way to offset the enormous gap in our energy budget.

Theoretical calculations of the vast savings from increased energy efficiency in industry are met by experienced operators with considerable skepticism. Those who have dealt with the nitty-gritty and know how reluctant Americans are to recede from levels of use to which they have become accustomed accept the more sanguine of these calculations at less than face value.

Desirable and indeed essential as conservation in the use of energy surely is, it is at best a short-term, in many cases a one-time remedy for energy needs in a country—or a world—in which energy needs grow inexorably, with the growth of population, with the rise of aspirations that call for more and more energy. Proponents of energy conservation who have put it forward as *the answer* to the "energy crisis" invite the charge that this is an "elitist" doctrine and a form of the "no growth" concept. Viewed as a short-term measure, as a spur to efficiency and a campaign against waste, conservation has everything to be said for it. But to offer it as a basic social doctrine of "less is better" or as a slogan to rally opposition to nuclear energy expansion and the excesses of industrialization and "consumerism" will antagonize rather than persuade the general American public and a world in which poverty closely connected with energy shortages is endemic.

Where does *nuclear energy* fit into the far from euphoric picture of future energy supply?

We have in this country seventy licensed nuclear power reactors in operation at forty-eight sites, with more than ninety additional plants in various stages of construction. In 1978, 12.8 percent of our electricity came from nuclear power; the most knowledgeable forecast for 1988 is 27.3 percent. The percentage is far higher—twice as much—in key industrial areas.

We rely heavily on nuclear power to keep our economy going.

For the near and long-term future, the energy we now have and can count on, from all sources, is not enough. Except for temporary periods it has never been enough, and it will never be enough for the kind of developing country we are, with our population steadily increasing and our desires and incomes expanding without long-term letup. I have listened for years to assertions that we don't need more energy; they have always and everywhere been wrong, and they are just as wrong today as they have been throughout the history of energy and industrialized economies. It is difficult for those who have never operated an electricity system to comprehend that the very availability of energy—especially electric energy—itself creates an ever expanding demand for more energy. The electrification of American farms, from virtually zero only a few years ago to almost 100 percent, is only one illustration of the many that come to mind. Today it would be unthinkable for American agriculture, both small and large, to have to get along with even a relatively minor curtailment of electricity supply or a slowing in the increased availability of electricity for farming and farm life. The public and political clamor that would result from a cutback of electricity to the millions of farms would be horrendous.

Energy is more than an impersonal statistic to be bandied

about by computers and theoreticians. Energy is part of a historic process, a substitute for the labor of human beings. As human aspirations develop, so does the demand for and use of energy grow and develop. This is the basic lesson of history.

America cannot presume to speak for other nations concerning energy needs. We are high energy users, and we should not be defensive about this fact. We may very well need to be serious now about moderating our consumption in the face of threatened temporary shortages and lowered demand due to our present sluggish economy. What is at stake for us is the maintenance of an already high living standard. For many other peoples energy sources are scarce, and therefore their living standards are low. Their need for more energy is desperate. Inevitably they look to us and to other highly industrialized nations, needing to use our technology to develop more abundant supplies of energy.

We and the rest of the world require vast and increasing amounts of energy, energy in all forms and from all sources.* We cannot count on any single source to provide the answer, or even any combination of sources, so fast does the political and technical picture change. We cannot reject out of hand any possibility, including nuclear power. But it must be safe.

---

*A plenary conference dealing largely with the world's needs for nuclear energy was held in Vienna in February 1980. The conference issued a communiqué on behalf of the sixty-six countries and five international organizations participating in a two-year study of the subject. The findings were summarized in the communiqué, as follows: nuclear energy is expected to increase its role in meeting the world's energy needs and can and should be widely available to that end; effective measures can and should be taken to meet the specific needs of developing countries in the peaceful uses of nuclear energy.

# 3

# MORATORIUM—AND A NEW START

How did we come to have fundamental and fear-ridden reservations about a major program that has been rushing ahead for more than twenty years, and upon which so many now depend in their daily lives and work?

The story is strange but understandable.

In the early 1960s American government and industry made a huge commitment to one—only one—method of converting nuclear fission into electricity. Other methods were proposed, but the alternatives were judged to be more costly and otherwise less acceptable by the responsible authorities—the Atomic Energy Commission, the congressional Joint Committee on Atomic Energy, and industry. To make atomic power a reality, and to do it fast, government and industry plowed ahead at a furious pace, having convinced themselves that this one method—the "light water" reactor—was the final answer. The prize was dazzling: a multi-billion-dollar national atomic energy industry with scores, even hundreds of plants planned throughout the country.

This proved to be a bonanza for American prestige, technology, and industry; American plants of the light water design preempted most of the European atomic market as well—and they still do. But the problems and the risks of this method were grossly misunderstood and underestimated in the '60s and '70s—and still are.

I do not want to minimize my own role in this inadequately considered rush to adopt the light water method of atomic energy production. When I was AEC chairman, I warned of radiation hazards and encouraged efforts to eliminate or reduce them, but the fact remains that I was the chief early promoter of atomic electricity. Moreover, I was largely responsible for interesting private investment in this field. I wanted peaceful atomic energy to become part of the mainstream of American life. I spoke out publicly and often in favor of private participation, arguing that a continuation of the government monopoly would have a limiting and constricting effect over the long term. One result of my advocacy of private involvement in atomic energy was to set the stage for the subsequent commitment to the light water fission conversion method, a method whose defects we were aware of but which, because of our inexperience with a new and complex technology, and in our haste, we failed to take into full and sobering account.

When I left the AEC in 1950, atomic power was still far from being a reality. The first prototype reactor, at Shippingport, Pennsylvania, managed for the AEC by the redoubtable Admiral Hyman Rickover, went into operation in 1957, and by 1962 four more plants were completed. Then came the boom years. By 1968 no fewer than seventy-five commercial nuclear power plants were in various stages of planning, design, and construction, and most of them were completed and in operation within the following decade.

In the early 1960s I became increasingly alarmed by this rush to throw up plants across the country without proper regard for safety considerations. I began speaking out about the risks and alternatives; in a way, I was a sort of early "no nukes" protester.

In 1963 I made a formal challenge to the good sense and safety of the atomic energy program when I delivered that year's Stafford Little lectures at Princeton, subsequently expanded in my book, *Change, Hope, and the Bomb*. (In a later writing I termed the "rash proliferation of atomic power plants [in this country] . . . one of the ugliest clouds overhanging America.") This public dissent to the "headlong" commitment to the light water reactor method from the first chairman of the AEC did not go unnoticed; I was given rough treatment by the governmental and scientific atomic establishment for my temerity.

There were few questioning voices to support my position back then, no Union of Concerned Scientists, no mass demonstrations of protest. Now the questions have multiplied, and the opposition is large, widespread, and vocal. The hurricane of protest that Three Mile Island evoked was no isolated scare reaction; it was the result of a long process.

The primary reason for my concern and criticism in the 1960s of the accepted American light water method of producing electricity from the fission of the atom centered on issues of safety to human health. But there was then and there continues to this day to be an even more serious hazard and fault inherent in this method, a hazard described by Senator Charles Percy as the "worldwide proliferation of the capacity to fashion deadly nuclear weapons from peaceful nuclear materials," a hazard which "relates directly to the future of our own and world civilization."

A simple statement of the physical facts about the prevailing method of producing atomic energy for peaceful civilian purposes makes plain the basis for this alarming crisis; it also provides a perhaps decisive reason why a different method of securing peaceful energy from the atom seems to me to be an urgent, top priority necessity.

The case can be put briefly: the act of producing electricity for peaceful purposes, under the present method, also produces plutonium, the basic ingredient in thousands of atomic warheads and an essential material for the making of additional atomic weapons the world over. The spent uranium fuel rods of an atomic energy plant can be reprocessed to extract plutonium from them. This means that reprocessing the spent fuel of peaceful atomic reactors sold or licensed by this country (a large number throughout the world) can multiply and has multiplied the number of countries that have the capability of producing atomic weapons.

In January 1976 I was invited to state my views on this issue before a committee of the Senate.* I took the position that this country should immediately stop the practice, widely followed by the AEC for years, of exporting to foreign countries the technology and equipment by which the spent fuel could be reprocessed and the bomb material—plutonium—recovered. I called for an embargo or a moratorium on these exports until effective international control was established. I testified that "many private citizens will be shocked and indignant when they come to recognize the extent to which the United States has been putting into the hands of our [domestic] commercial interests and of foreign countries quantities of bomb material; that we have been shipping this stuff all over the world for years." America, I said, is the world's leading proliferator of atomic weapons. I listed the large amounts of weapons material this country had been—quite legally and openly—providing to foreign countries, thereby stimulating the spread of nuclear weapons throughout the world chiefly under the attractive but misleading public relations slogan, Atoms for Peace.

The administration of President Ford subsequently agreed with my position as applied to the export of technology for the

*Hearings Before the Committee on Government Operations, United States Senate, Ninety-fourth Congress, second session, on S. 1439, January 19, 20, 29, 30, and March 9, 1976.

reprocessing of spent fuel. The Carter administration continued the embargo, and it is the policy of the United States at present. Neither of the decisions by the Ford and Carter administrations went as far as I thought—and still think—we should go, that is, to institute "a complete and unconditional embargo on the export of all nuclear devices and all nuclear material, and American know-how" and that we should "without delay proceed by lawful means to revoke existing licenses of American [designed] nuclear devices held by American manufacturers."

At the heart of the issue of proliferation is, again, the very nature and faulty functioning of the American method of producing peaceful atomic electricity. It makes urgent the development of a different way of producing electricity from the atom that does not have this alarming side effect.

There are other potential hazards, more immediate and more direct. These were dramatized by the Three Mile Island accident. I happen to think that some of these hazards—radiation pollution of the environment, for example—were in many instances grossly sensationalized and magnified out of all proportion to the truth (and on occasion with no relation whatever to the truth). Yet the hazards exist, and the authorities have been unable to alleviate the growing public anxiety about them. Consequently there has been a further, more widespread "no confidence" movement the world over against the dangers inherent in the American method, now almost universal, of using nuclear fission.

It is not just that the man in the street has lost confidence in the safety of atomic energy plants. The atomic establishment itself—the scientific and industrial community—does not show much confidence either, and that has accounted for much of the emotional tension surrounding this issue. The atomic energy people have found themselves forced to defend a system that they realize is not wholly dependable; yet their pride is invested in it, along with their careers, their reputations, and, in the case of industry, their capital. Under those conditions their voices sometimes become as shrill as those of the opposition, and they are just as prone to irrational appeals and public relations gimmicks.

Because of this growing controversy—and because of the rapidly rising costs—we find that the outlook for nuclear power has changed radically from the rosy picture of just a few years ago. Plants have been blocked and construction delayed or halted. Orders for new plants have fallen off; many have been canceled outright. Projections of just a few years ago, that there would be five hundred or even one thousand reactors in operation in the United States by the year 2000, have been revised dizzily downward. Atomic power is now on the defensive, it is in retreat, and one of history's most inspiring opportunities to provide amply the needs of an energy-poor world population is in danger.

Today there is in effect a moratorium on nuclear power.

The "reassessment" of the entire atomic program that I called for in 1963, which was rejected, with boos, by the scientific and industrial community, is all the more needed today. The time is ripe not only for a new look at the program but for a new start.

Let me emphasize the word *start*. Many critics and opponents of nuclear power say that we should simply stop, close down the plants, and forget the whole business. In fact, some proponents and defenders of nuclear power say what is in effect the same thing; that is, some of their proposals for ways to minimize hazards (to make the atom "acceptable," as they say) are so extreme and impractical that they amount to saying forget it.

We need to turn our backs on the past—but not to quit. Nuclear energy remains a fundamental discovery of incalculable present and potential benefit. One way was chosen—chosen too hastily, largely for short-run commercial reasons—and has been followed too frenetically; but this was not the only method then, and it is not the only possible method now.

The present de facto moratorium gives us the opportunity—and the necessity—to find a better way, to develop a less hazardous method of converting into electricity the fantastic energy that lies within the nucleus of the atom.

Our main objective in the past was to produce electricity with a cheap atom, the cheapest possible, one that would be economically and environmentally competitive with other forms of energy at a time when foreign oil was absurdly cheap and plentiful. This simple formula was shortsighted and dangerously wrong, as we have learned to our cost.

We have poured countless millions into the light water method. Its apparent economic advantages lured us into rushing into it, as though it were the best feasible answer, and downgrading with scant attention the other possibilities. Now it is time to recognize as a matter of national policy that the method is not good enough, not safe enough, not the right answer.

Our new objective must be to find a safer, "healthier" method of producing peaceful energy from the atom by a method that also minimizes or eliminates the present risk of furthering the spread of nuclear weapons. If it is also cheap, or relatively cheap, fine; but first it must be safe.

It is not clear nor, perhaps, is it likely that any of the other known methods—those using heavy water, helium, or molten salt, for example—would yield a safer answer. I certainly do not know. That is the point: *nobody* knows for sure. These and other possible methods, which were considered and discarded, have not been thoroughly and rigorously exposed, through full scientific investigation and testing, to the objective of greater safety. They may not be improvements as measured by that objective, but we must find out. The same uncertainty about safety exists for the several versions of a breeder reactor now delayed for reasons related to proliferation. Certainly other methods will be developed in the future. If the history of technology and science means anything, it means that the possibilities are limited only by the ingenuity and curiosity and genius of human minds.

If the entrenched scientific establishment wrings its hands and says that a completely safe atom is an impossibility—"against the laws of physics"—we should not let that impress us. As most of us have learned, establishment people in all walks of life are expert in the art of saying something can't be done. The lesson of such gloomy forecasts may be not that the thing cannot be done but that we need new scientists, new and fresh ideas from those not committed to existing methods, whether those conventional methods are scientific, technical, or financial. The noted scientist Freeman Dyson, in his recent book *Disturbing the Universe,* tells us a too familiar story, that of a promising scientific atomic adventure looking toward a safer atom thwarted, largely by conventional financial myopia and the boredom of worn-out geniuses.

And a new and perhaps younger generation of enthusiastic scientists and idea-men there will be, particularly if they are encouraged and funded by deliberate national policy to participate in the challenge for whatever span of time it may take—five years, ten, perhaps a generation—to find a safe and abundant supply of energy from the atom for all the peoples of the world.

An aura of resignation and fear hangs like a cloud over atomic energy today. In its proposed national energy program and in its 1980 budget message, the Carter administration virtually ignored nuclear energy, relegating it to the status of a last desperate resort in the event all else fails.

Yet the enormous potential of nuclear power remains; we simply have not yet found the right way to use it. Here I propose the rudiments of such a program, and I invite others to join in the hunt for a clean atom.

How should we organize the undertaking? What should our policies be? Perhaps the most that can be done at this juncture is to set up some criteria. To stimulate the discussion, I will suggest that Congress charter a corporation, one analogous perhaps to Comsat in the field of communications or to the National Science Foundation.

I suggest that this New Atom Corporation, with both private and public stockholders, be empowered to carry on the research and development of new methods of producing electricity from the energy of the atom.

The enterprise should be organizationally independent and immediately accountable to the President.

It should utilize existing research groups such as the national atomic energy laboratories, the universities, and private industrial research centers. But it should particularly seek to develop *new* research initiatives.

The basic policy of the New Atom Corporation should be to encourage diversity, a multiple approach. Its charter should give explicit recognition to the international nature of atomic energy and the energy needs of the people of poorer countries.

The New Atom Corporation should be considered part of the *total* national energy program. This will stimulate the cross-fertilization of ideas and concepts from research in other and diverse energy sources.

The reader may well ask, what about now? It's all very well to talk about some possible future discovery of relatively hazard-free atomic energy, but that does not answer the immediate critical needs and the practical questions of today, with almost a third of our electricity supplied by the atom. What do we do with the seventy operating reactors? And the ninety-odd reactors being planned or built? Do we close them all down? Must we put the atom on the shelf while we wait for a better, safer method?

My answer is an unqualified No. We desperately need the existing plants, and we should do everything possible to maintain them in operation, except in those cases where technical or operating defects create clearly hazardous conditions for employees and the general public. (The Nuclear Regulatory Commission has already taken some action along this line.) Notwithstanding Three Mile Island, the overall atomic safety record is good. The hazards exist, but for the present they appear to be acceptable when balanced against the quite different hazards that a nuclear energy ban would bring. The sudden unplugging of atomic power in Chicago would just about shut that giant city down, closing factories, darkening streets, and dimming lights in homes. Try to imagine Boston without a third of its electricity supply; or Providence, or Hartford; or the scores and scores of cities and towns throughout the Northeast and in many parts of the Midwest. Electricity means production, and production means jobs, income, security. Its loss could mean near chaos. A drastic cut in electricity would have an immediate, seriously damaging effect on our national life. The shutdown cure for atomic hazards could be worse than the disease.

The reality is that we are now so heavily dependent on atomic energy that we are going to have to put up with what we've got for some time to come.

As for the ninety or so reactors in various stages of planning and construction, it does not seem possible to formulate a general rule. They must be examined case by case. No doubt some of them should and will be completed and put in operation; others will be deferred or perhaps abandoned; and still others will be redesigned for greater safety on the basis of analyses of the Three Mile Island accident and other instances of technical and managerial malfunction. The NRC has been moving in this direction. The Tennessee Valley Authority, the nation's largest atomic power producer, has taken its own initiative to tighten up on safety.

In order to keep most present plants operating and to complete some projected new plants, however, there must be a fundamental change in attitude in the atomic industry. There must be no more purely lip service paid to safety, no more easy acceptance of radiation hazards. From now on the atomic power program must be carried out on a safety-first basis.

I propose, then, a double program: first, for the long range, a new start in nuclear research and development, aimed at finding and proving out safer and better methods; second, an immediate program to establish and enforce new and exacting standards of safety and system management, to provide the maximum possible protection against radiation hazards in the operation of the badly needed existing plants and those additional reactors that necessity and good sense may dictate that we complete.

# 4

# THE PERILS OF IGNORANCE

The general public needs to have some idea of what atomic energy is and how it works if we are going to have a sensible and democratic long-term program of nuclear development. Not a little of the argument and debate over atomic energy has been and continues to be marred by misinformation and confusion. This situation has in part been responsible for the tension and fear that surround the subject.

Much of the public's fear is misdirected. What are we really afraid of? We are afraid of the thousands of nuclear weapons in this country, in Russia, and the capability of making such weapons in a steadily increasing number of other nations (the list grows almost yearly), which together threaten a terminal catastrophe, even the extinction of life on this planet. *This* is the deep and real fear. Beside the looming spectre of atomic war, the hazards of nuclear electricity production are indeed minor. And yet the argument over power for civilian uses has suffered from a confusion with the threat of atomic weaponry—quite a different thing —and a heavy carry-over of emotion that has burdened and complicated an already sufficiently weighty and complex subject.

Sometimes this confusion of the issues seems deliberate, sometimes it is blatant sensationalism or ugly demagoguery. It is as if some nuclear opponents believe that abolishing the lesser evil

(radiation from malfunctioning nuclear plants) will somehow reduce the greater evil (the bomb). This exercise in futility illus-
trates how hopeless and helpless most people feel about the clear
and growing danger of atomic war. The nuclear arms race has
gotten so far out of control that, for many, it seems quite impossi-
ble to confront it with any emotion but despair or resignation.
Atomic electricity, on the other hand, is not yet beyond the reach
of corrective public action. As a surrogate for the bombs, atomic
energy can be fought, protested, and possibly defeated, even
though such a victory would not resolve the issue of the presence
of atomic weapons.

How can we substitute information about the atom for ignorance of it? This is no easy matter, as I have had occasion to know
personally. When I was AEC chairman in the late 1940s, I made
a big point of the necessity for public education in atomic energy.
I called for a large information program in the schools and the
mass media. On one occasion, after I had appealed to the press
and radio to undertake such an education program, my listeners
responded, We'll be glad to help, but just exactly what are we
supposed to educate the public about? Public education as a
concept sounded fine, but when it came down to cases, it wasn't
clear how much the public needed to know or how this body of
information was to be imparted and absorbed.

At that time atomic energy was not only a revolutionary new
force, it was an abstract thing that didn't really enter into people's
lives in any direct, tangible way. There weren't any handles on it.
After all, how many people know very much about the workings
of the cars they drive and depend on every day? Not many; yet
that doesn't keep them from having reasonably well informed
opinions on public issues involving automobiles (highway construction, gasoline taxes, etc.).

So our original impulse concerning public education in atomic energy has had to be tempered by a more realistic appraisal. We can't expect people to become physicists any more than we can expect them to become auto mechanics. However, they do need a certain fundamental grasp of what the atom is and how a fission reactor works in order to give some background to the common-sense judgments they must make about atomic energy policies. And we can expect a more thorough knowledge of atomic energy on the part of those who have special public responsibilities—public officials, political leaders, and press and television reporters and editors.

Some progress has been made since the early years. The public now has a general acquaintance with atomic energy, and the mass media and public officials are as a rule much better informed. But they are not yet well informed enough, as the Three Mile Island accident showed. The record of the press, and television in particular, in covering this major atomic story was spotty at best. Some reporting was well balanced and fairly accurate; some was full of error and verged on the cheapest kind of sensationalism. As a whole, the media showed a disturbing lack of knowledge. The workings of an atomic reactor have been no secret for nearly a generation, but from the way some of the press stumbled and scrambled, one would have thought that nuclear power had been discovered only the day before.

Reporters weren't the only ones to show themselves ill-acquainted with fundamental facts. Public officials, too, were to be found in that category, including some who have responsibilities in the field and who should have done their homework. One leading member of Congress, emerging from a nuclear plant inspection tour, said resignedly to the press, "It is just too complex to understand."

That is not so.

The engineering of the conventional light water reactor—the

pipes, the valves, the complicated gadgetry—is highly intricate, and no layman should be expected to understand a fraction of it. But that is not necessary. The basic principles are easy enough to grasp, and they have been described clearly in many school texts, popular magazines, and books. The nub of the matter can be stated quite briefly in a few paragraphs.

The *furnace* of an atomic reactor contains an assembly of slender tubes, or rods, of processed, slightly refined uranium. This is the fuel of a reactor. To "ignite" the assemblies, so that they will emit and sustain powerful radiation and the resultant intense heat, carefully calculated numbers of these fuel rods are brought together in the quantity necessary to initiate and sustain the continuous "chain reaction." It is the spontaneous splitting of atoms—fission—that creates radiation and heat. The intense heat causes the water circulating around the rods to become steam. That steam is fed into a turbine, which in turn drives an electric generator to produce electricity. The water is ordinary water; the term "light water reactor" distinguishes this method of cooling the core from the method in which a reactor is cooled or moderated by "heavy" water, water processed especially for use in a reactor.

The term *furnace,* as applied to the radioactive core of an atomic plant, may evoke an analogy with the burning of coal or oil—chemical combustion—beneath a conventional steam boiler. This would be misleading. The temperatures produced by atomic fission are fantastically higher; most metals exposed to chemical combustion—a steam boiler burning coal, for example—simply would not stand up to the heat of the fission process. This fact alone calls for an extraordinarily high degree of design skill and workmanship on every foot of piping and every instrument.

An even greater difference between a nuclear reaction and a chemical reaction is the furious and lethal radioactivity that is created within the reactor core by the splitting or fissioning of atoms. The uranium assemblies must be removed from time to time, and fresh fuel assemblies put in their place; the old fuel rods, still powerfully radioactive, must be handled by remote control and with the utmost care. (The storage and disposal of these radioactive "waste" materials is a serious and controversial problem that I shall discuss in a later chapter.)

There are three methods of moderating and controlling the fission process that takes place within the atomic reactor. First, and of least significance, is the presence of the zirconium cladding that sheaths the fuel rods themselves. Second is the insertion of special rods (of boron, for example) into the container of the active fuel assemblies, or the *core*. The special rods absorb neutrons and slow down or stop the process. Third, and the principal means of moderation and control of the radiation "fire," is the circulation of water. The technical devices used to carry water into the core area call for a high level of design and workmanship. These plumbing refinements, so vital to the control of the reactor and to its safe operation, are not of vital concern to the public; the important fact is that water, the "coolant," must always be kept flowing inside the reactor vessel in order to prevent the activated fuel rods from becoming overheated and damaged. This moderating operation is designed to be controlled by the plant technicians, who constantly monitor the reactor by means of instruments located outside the reactor container.

As an added measure of safety, the nuclear component of the plant is encased in an airtight containment building and then in a very heavy shell made of special steel and concrete, which prevents the escape into the atmosphere outside the plant of any radioactivity that might result from a malfunction. (The huge cooling towers, shaped like giant milk bottles, that thanks to television have become visual symbols of nuclear energy, have nothing whatever to do with nuclear activity. They are there to condense steam into water. The "plume" of vapor from such a cooling tower that so agitated some television commentators at Three Mile Island was steam, not radioactivity, and it was as harmless as the steam from a teakettle in your kitchen.)

The foregoing thumbnail description should corroborate my belief that people do not need to know much physics or engineering in order to give a little substance to their ordinary common sense about atomic energy issues and to allow them to distinguish what is relevant from what is not, the true from the false.

A recent example of the pitfalls of ignorance was a documentary by a national television network on possible links between A-bomb tests in the early 1950s in the southwestern United States and a rise in cancer deaths among the people living in that area. Television reporters gave the impression that this story was relevant to the present controversy over the safety of nuclear power plants, yet there was no connection in any meaningful sense. True, both the bomb tests and atomic electricity production involve radiation; but drawing an analogy on that basis is like drawing an analogy between airplane crashes and people slipping on icy sidewalks, simply because both involve the force of gravity. (Certainly no one would argue that an analysis of airplane crashes would shed light on how to make sidewalks safe in winter.)

If a major television network can make such a fundamental and glaring error some thirty-five years after the beginning of the so-called atomic age, the public (which depends on the media for much of its information) can certainly be forgiven its failure to know all it should about atomic energy.

# 5

## THE ATOMIC ESTABLISHMENT

Why does the ordinary citizen need to know anything at all about atomic energy? Why can't one leave the entire business in the hands of the scientists and engineers and technical and managerial specialists whose job it has been for more than twenty years to design and build and operate nuclear plants?

The answer is because these experts are no more infallible than any other experts—which is to say that they are very fallible indeed—and they need to be subjected to the checks and balances of cross-examination and inquiry that are at the heart of the democratic process in an open society.

Unfortunately, this process has not been greatly evident in the brief but important opening years of the atomic age.

I would put it even more strongly than that.

The greatest hazard of atomic energy is not the faulty engineering design of plants or their poorly managed operation (though both were painfully apparent in the harrowing experience of Three Mile Island). The greatest hazard has been the history of ignorance of and complacency about these problems on the part of scientists, engineers, manufacturers, and government authorities. And not only their complacency, but their positive disdain of critics and questioners.

I do not exempt myself from some responsibility for the short-

comings with respect to safety measures. In the earliest days of
the AEC it was our urgent duty to build up a stockpile of bombs
when we had none. Under this highest priority rush, time did not
permit our giving adequate attention to safety questions. There
were gaps then in our knowledge about radiation hazards, and
there still are gaps now. We lacked experience. We made mistakes, as judged by today's standards and knowledge. For example,
we continued the practice (inherited from the Manhattan Project) of putting highly poisonous wastes from the plutonium plant
at Hanford in Washington State into steel casks that were then
buried in an uninhabited area of desert. These casks, after three
decades, are now beginning to leak, though not dangerously.

I don't want to neglect the military, either, in this catalog of
those responsible for the ignorance and laxity concerning the
hazards that have plagued the atomic program. I am writing now
about civilian atomic energy, but one cannot make a clean and
neat historical separation. The development of atomic energy
began as a military program—the bomb—and the course of subsequent events (and the attitude of Congress, scientists, and the
various administrations) has been heavily influenced by the military program of bomb making and aboveground testing. And the
military has displayed its share of complacency about the hazards,
a complacency bordering on recklessness. After all, the military
is more sheltered from public accounting; it has a powerful lobby
behind it and a strong and admiring and largely uncritical media
backing. Civilian officials and the media are rarely eager to confront it with charges of having made mistakes—yet mistakes there
have certainly been.

In the early days of the AEC the military kept pushing us to
hold A-bomb tests (for quite legitimate reasons of planning and
research and product testing). We insisted that such tests be
carried out in remote parts of the Pacific. After I left the AEC,

the military and scientists in the weapons program pushed for testing in the United States itself. Being closer to our domestic weapons research centers, such tests would provide quicker and cheaper results important to the design of "better" atomic weapons. Accordingly, the test series was held in the deserts of the southwestern United States, with soldiers in attendance and exposed to fallout—whose potential aftereffects on their health have only in recent years begun to come to light, largely through the diligence of a senatorial staff investigation.

One notorious instance of official cover-up had as its tragic centerpiece a Japanese fishing boat, the *Lucky Dragon*. It was elaborately, officially, and falsely denied by the AEC that miscalculated fallout from an American test in the Pacific had descended on this vessel, having a deadly effect on its crew. The subsequent public exposure of this deception had a damaging effect on the international reputation of our government and on that of the AEC authorities as well.

All these things took place along with the growth of an atomic consortium, that is, groups of bureaucracies in private and public life that have a vested career interest in the status quo and that guard their reputations and privileges as zealously as any other "establishment." With this difference, the atomic establishment, to reinforce its papa-knows-best tendencies, had the added thrust of the official mystery and complexity of atomic energy. They were the experts; they knew it all; it was over the head of the public, and the public critics were viewed (in the contemptuous words of one AEC official not too many years ago) as a "bunch of housewives."

Today, because of the vigor and determination of the antinuclear citizen protest movement and the merit of some of its objections, there are those who would write off atomic energy altogether. To accept this conclusion as foregone is to misperceive the strength of the other side, that of the groups who support nuclear power. The atomic establishment represents a broad spectrum of influential and responsible citizens and interests in many countries throughout the world. The financial stake in the existing atomic power industry is enormous. Many millions of dollars have been invested in what was until recently viewed as a "growth" enterprise whose future (and profits) had virtually no limit. Some of the nation's most potent corporations have expectations in this field, both in the United States and, notably, abroad, that they will not simply sit back and watch evaporate. The world prestige of these companies is of no small importance to them, and that prestige is involved here. Many other kinds of organizations have important interests in rebuilding the flagging impetus of the atomic power programs. These include publicly owned regional development agencies such as the TVA and its 160 rural and municipally owned electricity distributors; scores of scientific and technical communities; great universities, representing some of the world's most gifted and dedicated minds; and the federal government itself, with career people in the agencies concerned and on the staffs of the great national laboratories.

It is a formidable army that marches under the banner of *status quo atomicus.* If the controversy over the future of nuclear energy were to be decided by a trial of political strength, there would be no contest. The pronuclear interests are immeasurably the more powerful. The protest marches, sit-ins, and environmental lawsuits are a nuisance to them, but once they rally their forces, there will be little doubt as to the outcome.

I hope that the question will be settled in a more rational way; over time I believe it will be so settled, on the basis of a sober analysis of the facts, our national energy needs, and our international responsibilities. (I certainly do not want to make a blanket condemnation of the posture and attitudes of all within the atomic establishment. There have been those who have genuinely sought to bring errors to light, to correct them, to welcome criticism, and to listen to it.)

The effect of the general attitude of the establishment has been to fuel the forces of opposition. The frustration, the angry temper, and the political rhetoric of some of those now protesting—in some cases violently—against atomic energy are a reaction not so much to questions of fact but to the arrogant self-adulation of atomic experts and the open derision of the atomic equipment vendors.

I had to listen to their disdain more than once, when I was AEC chairman and found public-be-damned attitudes in the then youthful organization, and later, when as a private citizen and critic I clashed publicly with "official" atomic doctrines.

I will cite a case in 1963, long before anyone dreamed of putting a "no nukes" sticker on a car bumper, when I protested a nuclear plant project whose potential hazards, viewed in today's light, would make a bedtime story out of the worst imaginings evoked by Three Mile Island.

In December 1962 the Consolidated Edison Company, which provides the New York City area with electricity, formally proposed the building of a large atomic energy plant within the borough of Queens, a part of New York City and one of the most densely populated areas of the country. Con Ed's plans called for the installation of atomic reactors to be manufactured by the firm of Babcock & Wilcox on a design of Westinghouse Electric Co. At that time the Atomic Energy Commission and the Congressional Joint Committee on Atomic Energy were promoting a national program for a greatly expanded series of just such plants throughout the country; the Con Ed management's decision was prompted by official and scientific optimism about the safety of atomic energy. And the utility's management was warmly commended for its initiative. No protests were reported from residents of the city.

Testifying before the Joint Committee on Atomic Energy in April 1963, Con Ed Chairman Harland C. Forbes projected the giant electric utility's confidence:

> Certainly we would never have filed the application [for a construction permit] if we had not been convinced that such a plant can be safely operated. . . . We think it represents the ultimate in safety. . . . There was absolutely no public opposition to the [atomic] plant we built at Indian Point [thirty-six miles north of New York]. . . . It seems to me that the public has reached the point where they have accepted nuclear plants as a matter of course as they would any other plants. . . . I don't think we have had more than half a dozen letters about it. There have been a few community meetings at which one or two people have raised some question about the genetic effects of radiation and so forth, some of which is rather silly.

As a private citizen with no professional connection or financial interest of any kind with atomic energy since I resigned as AEC chairman, I was by no means persuaded of the safety of this proposed plant in a densely populated area. In public lectures at Princeton University in February 1963, and in a book based on those lectures, I questioned strongly the safety of such a plant and the measures available for the safe disposal of the radioactive spent fuel.

This criticism angered members of the Joint Committee on Atomic Energy, and I was summarily ordered to appear before it. There followed a lively session of hostile questions and aspersions from committee members. Excerpts from the 1963 hearing will indicate how important it is that the skeptical layman be persuaded by *facts* concerning the performance of atomic energy plants, not by assertions about atomic energy made by "experts," however prestigious they regard their own qualifications.

The *fact* is that neither the regulatory authority (the AEC) nor the most influential physicists nor the operating utilities nor the new profession of nuclear engineering, showed clear signs of comprehending the fundamental forces they were dealing with in the design and operation of reactors intended to control and utilize massive radiation and energy of magnitudes previously unheard of. It is dismaying to have to admit that what was true, as I saw it, in 1963 was still true in March 1979, at the time of the Three Mile Island debacle. And it is largely still true.

For the light it throws on the atomic policy stalemate of the 1980s, I quote here from my 1963 testimony before the congressional committee.

REPRESENTATIVE HOSMER: Let me ask you specifically with respect to the Con Edison plant that you use as an example. Are you fully familiar with the proposal respecting engineered safeguards against radiation and the reactor hazards which have been proposed with respect to this particular plant at its particular location?

MR. LILIENTHAL: No, I am not.

REPRESENTATIVE HOSMER: You charge, however, that you would not dream of living in Queens if that plant existed there.

MR. LILIENTHAL: Yes. I will give you my reason. It is not on the basis of what the engineers have to say. It is on the basis of thirty years of experience with engineers and scientists whose prescience and knowledge upon subjects frequently turn out to be less than their own estimate of their prescience, or their own view of what they say they know. . . . I am speaking from a long conviction and experience with technical people [when

I say] that they are human beings who are fallible. When it comes to safety, I take these predictions with a grain of salt. . . . In other words, engineers are fallible human beings. I myself would not want to live around one of these plants.

REPRESENTATIVE HOSMER: Can you honestly tell us that you are capable of assessing the safety or non-safety of this particular proposal in such a way as to feel with assurance that you are telling either us or the rest of the American people what the actual facts are?

MR. LILIENTHAL: No; I have said I don't have detailed knowledge. I think one of the best ways to get confused is to get too much detailed knowledge. . . . I would only suggest, if I may, to you and to the engineers for the Consolidated Edison, [that] they better make [the unresolved dangers] awfully plain to the people of Queens before they build that plant, because there is a skepticism among us ordinary laymen about the human fallibility of the scientific and technical mind.

My skepticism was certainly not shared by the committee in general. The tenor of its views was summarized by a comment of its chairman made to the chairman of Con Ed, the applicant for a construction permit.

CHAIRMAN PASTORE: Mr. Forbes, you and your company are to be congratulated, indeed, because you will be pioneers in [the] establishing of a large reactor in a densely populated section.

So much for congressional complacency about hazards.

Further confirmation came a few days after my testimony before the joint committee. An entry in my journals for July 25, 1963, gives a sobering description of the ignorance, or cover-up, on the part of the commission itself.

The issue before the Committee (presented by a member of the AEC, Dr. Robert E. Wilson) was an item of $19 million for a "standard pressure shell that you see all over the country." In this shell it is proposed to duplicate "the maximum credible [nuclear] accident, and find what the causes are."

The colloquy between the AEC experts and the Committee members confirmed my hunch that these experts and Con Ed's people have not told the public, or the Joint Committee, the whole story about the safety of such a plant in the heart of the city because they *just plain don't yet know* the causes that might produce a nuclear accident, nor just what damage such a "credible" accident would entail.

Here is one excerpt from the questions to Dr. Wilson and the testimony that runs several pages.

REPRESENTATIVE WESTLAND: How have you got your siting criteria now? We had Consolidated Edison appear before us the other day, and they indicated that the containment building that they were proposing, which is to be located in a very highly populated place, was going to be adequate.

Now you come in and say you are going to blow up some of these things in order to provide information or criteria.

The two do not seem consistent to me.

DR. WILSON: Do you not think we ought to know something about things like that?

REPRESENTATIVE WESTLAND: Sure I do. But I would assume you would know something about it now, or you would not be telling Consolidated Edison—you have not actually told them, but you have indicated you have told them—that it was going to be all right.

DR. WILSON: No, we did not indicate anything of the kind.

REPRESENTATIVE WESTLAND: I think Lilienthal was right, then.

Not long after I was given this going-over by the congressional committee, the AEC chairman, Dr. Glenn Seaborg—one of the world's greatest atomic scientists—rebutted my criticism concerning safety, declaring it to be an appeal to "unreasoning fear." With Con Ed's application to build an atomic plant in New York City pending before the AEC, Dr. Seaborg in a formal public address used the classic phrases. "I would live next door to the atom. I would not fear having my family residence within the vicinity of a modern power reactor built and operated under our regulations and controls."

Public opposition to the proposed Queens plant was beginning to develop, however, and some time thereafter the project was shelved. Subsequently, no atomic energy plant has been built within New York City, nor is there a plant within the confines of any major city in the United States.

The Con Ed episode happened to resolve around the question of nuclear plant *location.* My purpose in recalling it, however, does not concern that important issue (which I will deal with in a later chapter); rather, I cite it as an example of the sloppiness and arrogance that is inherent in a technical bureaucracy that goes unchecked by rigorous examination and hard insistent questioning from a determined, informed, and skeptical public.

A short time after AEC Chairman Seaborg gave his endorsement to the proposed Queens plant, I was the guest speaker at a crowded meeting of the American Nuclear Society in the ballroom of the Hilton Hotel in New York City. The audience included most of the country's leading nuclear scientists and engineers, representatives of the atomic equipment industry, and chief government officials.

I renewed my earlier expressions of skepticism about the adequacy of atomic plant safeguards and the progress—or lack of it—in confronting the public safety hazards of the radioactive spent fuel or waste from existing plants. I called for a serious reexamination of the entire atomic energy program, and I asked the members of the American Nuclear Society and their guests in government and industry this question: "What urgent necessity is there now that would justify taking any chances whatever with the still [as of 1963] unforeseeable consequences of miscalculation or human error or violation of the AEC regulations, that might possibly cause a major nuclear mishap in so densely populated an area as this or in the handling of the radioactive waste in a technology not fully tested?" I declared that I saw no such necessity for haste, expressed confidence in coal as a source of energy, and proposed that the vast federal expenditures on atomic energy be drastically curtailed while solutions to the hazards were reached.

On adjournment of the meeting, an AEC commissioner called a rump press conference in the lobby, and AEC employees, referring to themselves as a "truth squad," set out to rebut my speech to the members of the press who were present.

The rejection of serious criticism and skepticism could hardly have been more pronounced.

In the light of the widespread, worldwide decline today in public esteem for science and technology generally, I recall the opening statement of that American Nuclear Society speech of 1963.

So recently at the very apex of adulation and awe, with limitless financial support from the taxpayers, science and technical men today stand in danger of being exposed to the other extreme: growing disillusion on the part of the public, an avalanche of congressional faultfinding, accusations of confusion as to goals and methods, futility on a grand scale, and general disappointment in the proven results of huge federal expenditures.

These are portents you here would do well not to ignore, lest science and technology become a public scapegoat, as Big Business once was made a scapegoat, as bankers once were, and politicians. Such a wave of disillusion could be injurious to the atmosphere in which all intellectual activities flourish, injurious to the posture of science, both basic and applied, to the standing and support of universities and colleges. Nothing could feed a general anti-intellectual spirit more than the signs I see of a growing development of antagonism and distrust of science and technology.

My call for a reassessment of the atomic energy program—the theme of my remarks to the American Nuclear Society—came at a time when the regulators of the atom, the Atomic Energy Commission, were enthusiastically promoting a major, headlong expansion of nuclear plants throughout the country, with only a rhetorical gesture toward safety.

That was in 1963.

Things have changed since then, generally for the better. In the intervening years considerable technical and scientific advances have been made and experience accumulated, which make the safety outlook better. Better, but still not good enough. Atomic energy as a public issue has heated up to the boiling point, and the ordinary citizen hardly knows what or whom to believe.

The louder the contending voices, the more rigid the opposing positions become—and the more perplexed the public. What everyone needs is not a "victory" for the pronuclear or antinuclear forces but a common-sense approach that will benefit the nation as a whole—a victory, that is, for the kind of democratic pragmatism that has served us well throughout the two centuries of the American experience.

What I propose in this book will, I hope, fit under the heading of pragmatism. I am a firm believer in and a practitioner of the art of the possible. I have no great use for absolutes. I prefer taking one step at a time, a technique that I have called elsewhere "the manageable job."

The things I have to say will not satisfy the fiercer partisans on either side of the nuclear argument. I certainly do not join the more fervent antinukes by insisting that we close down our existing atomic plants; nor are my comments on the present system calculated to win applause from the hard-liners in the atomic establishment. Of course I hope that these passionate contenders will listen to what I have to say, but most of all, I hope the general public will find that my suggestions make sense—pragmatic sense.

When I urge the necessity of finding a safer method of producing nuclear power, I know that this almost certainly cannot happen overnight. Nor, in all probability, will it come from one of those strokes of genius—or inspired luck—that illuminate the pages of scientific history, including the saga of atomic discoveries. More likely this objective will be reached as the result of the patient accrual of small advances, the work of a great many men and women over what may turn out to be a fairly considerable period of time.

Likewise, my other major point—that we retain most of our present atomic plants but run them on a more rigorous, safety-first basis—is, I believe, a reasonable suggestion. The present hazards can be dramatically reduced if we place safety and good management ahead of a narrow concept of "cost-effectiveness"; it will not be easy, but it can be done—in the same steady, deliberate, and painstaking way that the search for a safe atom should be undertaken.

To reach this goal will not and should not require a radical reorganization and restructuring of our nationwide electricity enterprises, with an unsettling effect on the supply of electricity— a basic service to industry, homes, and farming. Nor should it require the virtual takeover of electricity management under the guise of federal "regulation," as is now being proposed by inexperienced but articulate novices in the business of electricity supply.

All this will demand that we keep our minds firmly set on our ultimate objectives while retaining the flexibility to accept something less than total success for the time being, knowing that each step forward brings us another step that much closer.

I find it ironic that we cannot count on much leadership here from the scientific and engineering fraternity. With a few outstanding exceptions, they who should already be in the forefront of an effort to achieve a better nuclear answer seem to have lost their nerve over atomic energy. The concept of the manageable job, so familiar to those of us who have had to deal with the imperfect realities of public and private business, appears not to be compatible with the thinking processes of most scientific and technical people confronted with the problems of public disapproval.

This may seem to be a harsh judgment on the part of one who as a layman has worked much of his adult life with scientists and technicians. It is based on what I have seen and what I know about how, in the universities and the laboratories, scientists and engineers are trained to think. (It applies as well to others, notably theoretical economists and computer-model zealots.) Their approach to problems is based on what they choose to call the scientific method, a method that is very exacting, and very precise, and in the end recognizes one and only one answer, the "truth" of the problem. The minds produced by this kind of training and outlook are uniquely suited to deal with many important and complicated questions. But not all questions; particularly not questions that involve human judgments and human passions, which are often shifting and willful, inconsistent, and at times contradictory. Confronted with such chaotic challenges, the man with the scientific mind will usually propose what to him appears to be the only solution. When that solution is not acceptable, or falls to pieces, he throws up his hands in despair and retreats, saying that the issue is a "political" one, meaning a *human* one, intractable to scientific thinking.

I have seen this happen time and again. After the horrors of Hiroshima and Nagasaki, most of the gifted scientists who made those grim events possible began calling for the immediate establishment of a world government as the one and only way of coping with the clearly perceived dangers of international atomic arms rivalry—the insane arms race that we are witnessing today. Nothing less than world government would do. Those who suggested it might have done better to negotiate concrete, specific, practical *beginnings* rather than concentrate exclusively on ultimate goals were impatiently dismissed. It was all or nothing; it became nothing. Many of these sincere and intelligent individuals, seeing that their scientific solution to the problem of planetary survival had come to naught, withdrew in puzzled dignity from the field.

Over the years I have found myself increasingly skeptical about the practical wisdom and staying power of scientists in the quite different but related area of domestic atomic energy. They are good at computing abstractions, such as the statistical probabilities of reactor accidents, but let a real accident like Three Mile Island come along, and they have little of a creative nature to offer. On pollution and other environmental questions they have shown a similar inflexibility and lack of ordinary understanding, and here they have received a clobbering from tough-minded and politically aware environmental groups and from equally tough industrialists. Events in recent years have taken the spunk out of much of the scientific community. One of my purposes in writing this account of my life with the atom is to encourage a revival of its positive, affirmative fighting spirit.

I find it dismaying that those individuals and institutions who are in the best position to know atomic energy's potential for humankind should adopt, in the face of criticism and setbacks, a posture of defensiveness and retreat. This is not humility; it is abdication. Positive steps need to be taken, positive things must be done. Unfortunately, those who by virtue of their expertise and their training should take the lead are not fulfilling their role.

If an affirmative spirit of venturesomeness for a new and better atomic energy program will not come from the established scientific and technical community, from where can we expect it to come? I suggest that it may well come from citizens without technical training. For they are fast becoming aware that they are the most concerned of all about finding a safe and a plentiful new source of energy.

It is altogether fitting that this should be so. From the very opening of the atomic age, more than thirty years ago, there were wise people who were also scientists, such as Dr. Henry Smyth, who expressed, in the very first report to the people of the world about these new discoveries, a conviction that I share: that the ultimate decisions about atomic energy must be made not by scientists but by those who in our society decide all major public issues.

# 6

## THREE MILE ISLAND

The Three Mile Island accident of March 28, 1979, was a fearsome demonstration of the fundamental truth about nuclear energy, that it is not just a different way of generating electric power but a basic force that, if misunderstood, underestimated, or mishandled, can bring disaster. Luckily, Three Mile Island was not a disaster. No one was killed or injured, and the radiation that did escape into the atmosphere was subsequently judged to be of no significant hazard to public health. The maximum radiation dose in the immediate area, according to the Nuclear Regulatory Commission, was 80 millirems (the average American absorbs 200 millirems of radiation from all sources during one year). No hazardous radiation was detected in food, milk, drinking water, or the river. The traces of radioactive iodine 131 found in local milk supplies were below levels measured after the 1976 nuclear bomb test by China—levels which were found to be insignificant at the time.

Still, in those panic-ridden early days of the accident, thousands of persons fled their homes. Many thousands more prepared for evacuation while state, federal, and local authorities, wholly unprepared, hastily threw together emergency plans. The nation—and the world—watched the drama of Three Mile Island unfold in an atmosphere of anxiety fed by scare

stories and careless inaccuracies in some of the press and television coverage.

What actually happened at Unit II of the Three Mile Island
plant?

The accident began with a malfunction of a pump that was
feeding water to the steam generators. Then a series of things
went wrong with other pumps and valves. Auxiliary pumps
switched on, as they were designed to do, but they could not
pump the water because, through negligence, their valves had
been shut down some time earlier and not reopened. Consequently the water backed up inside a secondary loop, and pressure
rose inside the reactor. This in turn caused a relief valve to open,
and it became stuck in the open position. The subsequent drop
in pressure activated the emergency core cooling system; water
poured into the reactor—and out through the open relief valve.
At the same time, malfunctioning instruments gave reactor operators erroneous readings, causing them to believe that the reactor
core was completely and safely covered with water coolant when
in fact it was not. Consequently they switched off the emergency
system on the assumption that it was no longer needed. This
premature shutdown caused the reactor's fuel rods to overheat,
reaching a dangerous temperature of about 2,500 degrees Fahren-
heit. Because of this heat the fuel rods suffered some local damage, but there was no "meltdown" (a word that is now a part of
the language).

The overflow of coolant water formed a small lake on the floor
of the reactor building. Some of this water was pumped into an
auxiliary structure that was not designed to handle and contain
high-level radioactivity, and gases given off by the radioactive
products in the water were picked up by the ventilation system
and carried outside, into the atmosphere.

The reactor itself had been shut off by an automatic safety

device (boron control rods dropped into the core, cutting off the fission process), but the reactor went on producing residual heat of tremendous intensity. A space formed in the upper part of the steel housing of the reactor core. It was feared (quite mistakenly) that this space might fill with enough gases to block the flow of coolant, thereby driving the core toward the nightmare region of a meltdown—or that the gas in the space (described incorrectly in early accounts as a hydrogen bubble) might explode, rupturing the walls of the container structure and pouring clouds of radioactive gas and steam out over the immediate area. Fortunately, there was no explosive mixture in the gases that collected, so these conjectures remained in the realm of hypothesis. In the tense hours and days following the accident, nuclear engineers and safety experts were able, through a cautious and skillful series of operations, to bring Unit II back under control.

A special presidential commission was set up to investigate the accident and determine with precision what human and technical shortcomings had brought it about. Yet the main lines of the Three Mile Island story were already clear enough. Over and above such specific causes as faulty valves and insufficiently trained control technicians lay the same attitude of complacency, indeed carelessness, toward the vital question of safety that has plagued the atomic power program since its inception.

I will cite as an instance of this the admissions made by officials of the company that built the reactor, Babcock & Wilcox, in testimony before the presidental commission in July, as reported by the *New York Times*. These officers conceded that the company had failed to take proper heed of warnings made the year before that B&W reactors might be liable to the kind of accident that subsequently hit the Three Mile Island plant. These warnings, made by B&W engineers, were based on a study of a reactor malfunction at a Toledo atomic plant. There, as later at

Three Mile Island, a pressure release valve stuck in the open position, threatening loss of the coolant around the nuclear core. Special safety pumps went into operation automatically, but— again, as happened at Three Mile Island—they were shut off by technical personnel when a faulty meter indicated wrongly that they were not needed. The pumps were restarted shortly there- after when operators realized that the meter reading was incor- rect. (At Three Mile Island this error went undetected for a considerably longer time; by then part of the reactor core had been deprived of coolant and was overheating.)

The reports on the Toledo incident proposed that operators of B&W reactors be trained in handling such emergencies, but nothing was done. A top B&W official conceded, "I wish we had acted sooner. A memorandum that lies around for six months or so is not something that I wish to repeat."

Babcock & Wilcox should by no means be asked to bear the entire burden for what happened at Three Mile Island. Others had important responsibilities for the project in its construction, operation, and supervisory control. The plant was a new one, perhaps a little too new; critics charged that it had been rushed hastily into year-end operation on December 30, 1978, in order to qualify for substantial tax credits. Then it had to be shut down for nearly two weeks in mid-January when two safety valves rup- tured during a test. Other technical "bugs" had to be eliminated in subsequent operations. During the crucial early hours of the accident itself, it became clear that plant operators had not been thoroughly trained and that emergency procedures, which in- volved the Nuclear Regulatory Commission and state and local authorities, were inadequate.

Three Mile Island offers a dire warning of what can conceiva- bly happen when safety considerations are not primary. Design errors go uncorrected until it is too late; operator training deficiencies aren't detected in time; technical and management faults somehow slip through the multiple nets of supervision, inspection, and regulation, and vital reports gather dust in somebody's in-box.

The major lesson of Three Mile Island is that safety can no longer be regarded as a matter of routine importance, along with such other important "technical" matters as efficiency and productivity.

In nuclear plants safety must come *first.* And it must *stay* first.

No discussion of what went wrong on Three Mile Island would be complete without reference to what went right. This was summarized by a knowledgeable practical engineer, Dr. D. B. Trauger, in a statement on June 4, 1979.

> No fuel melting occurred, even though the reactor core was uncovered. The safety system functioned reliably. The containment [into which cooling water poured in quantities] functioned effectively. Occupational radiation exposures were within annual limits. The maximum hypothetical public individual exposure was less than 100 millirems, which is comparable to a few months of living in the . . . environment of Denver.

The reactor was shut off; technical resources were available and were used effectively to reestablish control of the overheated core after many hours of patient and difficult labor.

It should also be noted that there have been no deaths in the more than twenty years of commercial atomic energy plant operations. (Three technicians were killed in the explosion of an experimental reactor in 1961 at the government's Idaho Falls test area.)

Nevertheless, some accidents—serious ones—have taken place. One of the most significant, and one still taken too lightly, occurred in 1966 when the core of the Enrico Fermi reactor near Detroit partly melted when a cooling system jammed. It was four years before the plant resumed operations; in 1972 it was closed down permanently. In 1975 electrical insulation in the TVA Browns Ferry reactor in Alabama was seriously damaged by a fire caused by an incredible piece of negligence, a technician testing equipment for air leaks with a lighted candle.

However, judged by the standards of other industrial operations over a comparable period of time, the safety record is excellent. Yet the point cannot be made too strongly: ordinary standards are not good enough for nuclear power. The stakes are too high; the risks are too great.

Three Mile Island is proof of that.

The question raised by the worldwide trauma following the accident at Three Mile Island goes far deeper than a consideration of whether the safety system of this particular method of producing electricity from atomic fission narrowly averted a disaster. The question the accident raises is whether this method is a reliable one for the long term. It is the method that (with minor variations) is followed by all of the seventy reactors now in operation in the United States, the ninety-odd now in development, and virtually all of the many operating reactors in Europe based on this same American technology and design.

My conclusion is that the prevailing method of controlling massive radiation is far too complex to serve as a reliable long-term basis for atomic energy. Such complexity is characteristic of an immature, first-stage technology. In light of the early stage of development, these reactors and their "plumbing" are certainly ingenious, but they are not good enough for a satisfactory, mature, and reliable technology. There are just too many junctures at which something can go wrong, too much monitoring and instrumentation required (all of which must always perform perfectly), too many possibilities for minor mishaps that would create public anxiety and unease.

It has become popular in the proliferation of second-guessing and multiple investigations concerning "what went wrong" at Three Mile Island to attribute the major troubles to "human error." But it is the designers' responsibility to design such a plant so as to take into account the likelihood—even the certainty—of human error. The plant should be designed to be operated by human beings, who no matter how well trained and highly motivated will remain human beings, ever subject to error. An overly complex plant, such as that at Three Mile Island, invites human errors.

A major, all-out technical research and development program for a *different, better, simpler* method is a necessity, and it cannot be started too soon.

# 7

# TVA: A LESSON IN NUCLEAR
# MANAGEMENT

For its nuclear energy for the present and the early future, this country depends on seventy reactors. Together they supply a substantial percentage of the total electrical energy generation in some of the most populous regions of the country, such as the Northeast, where not less than a third of the total electricity generation is presently nuclear, and the Chicago area, where it is more nearly 50 percent. This generation comes from plants of basically the same design as the plant that malfunctioned at Three Mile Island. A new, different, and safer alternative method along lines I discuss in this book can hardly be widely available in less than ten or, more likely, fifteen years.

This being so, for the immediate future we must make the best use we can of the present reactor system. And we should make very certain indeed that whatever their design shortcomings, the present reactors are operated on a safe basis.

The Tennessee Valley Authority provides an outstanding example of atomic power management and operation. This regional agency is the nation's largest producer of atomic power as well as its single largest power system overall (including major hydroelectric and coal-based power plants). TVA general management has had decades of experience, and its energy expertise is seasoned and capable.

Shortly after the Three Mile Island accident occurred I was invited to visit the Tennessee Valley to join in a celebration of the forty-sixth anniversary of the signing of the TVA Act by President Roosevelt. This weeklong visit took my wife and myself through the valley—from Alabama to Virginia—where, as a founding director of TVA and chairman I served from 1933 until I left to become chairman of the Atomic Energy Commission in 1947.

As early as 1950 TVA's technical forces began to familiarize themselves with the experimental work on peaceful uses of the atom being carried out at the nearby AEC complex at Oak Ridge, Tennessee. In 1974 TVA's first atomic energy plant went into service at Browns Ferry, near Decatur in northern Alabama. Now, in 1979, at a site in eastern Tennessee in the vicinity of Chattanooga, the second TVA atomic plant, Sequoyah, had been completed, and the huge enterprise was ready to be loaded with nuclear fuel and put into operation as part of the TVA's valley-wide electricity generating system. I was invited to examine the Sequoyah plant in detail, including the training program for operators, safety measures, and area environmental monitoring systems, to become acquainted with members of the staff, some of them TVA veterans, and to spend time with the large design organization at work on the six new atomic reactors authorized and in various stages of completion—the largest program for new reactors in this country.

In view of my freely expressed skepticism that safety had not been dealt with seriously enough in the industry generally, my special interests in this inspection were safety of design, the caliber and experience of management, and especially the program for the selection and training of operators.

My visit had been made open to the press and television, and I was accompanied on the inspection tour by two score reporters and photographers, who were free to ask questions and take pictures of any part of the plant. Therefore, what I saw and heard could be reported to the general public without restriction. I gave TVA a good mark for this openness, particularly as it was right after the Three Mile Island accident, a time when any nuclear power producer might have been less than eager to welcome critical attention from the public.

What I found at TVA encouraged me to believe that, imperfect and risky as it is, the prevailing design of light water nuclear reactor can be redesigned for improved safety and, if operated under more stringent codes, can provide a reliable stopgap energy supply until a wholly new and better method can be developed.

TVA technicians I found hard at work on improving design features, both as a regular part of their job and in response to early information they had received of what went wrong at Three Mile Island a few weeks earlier. The TVA people had not waited for someone to hand them a lengthy report; they had gone into action promptly and responsibly, and their initiatives and corrective ideas have been made available to other atomic energy utilities.

What impressed me most at Sequoyah was the human side of the enterprise, the selection and training of atomic plant operators, a process that extends over a period of several years and includes psychological and intelligence testing and periodic retesting.

The typical TVA engineer in charge of a "shift" must have eight to ten years of training and experience. Trainees must at least be high school graduates, and they must spend sixteen months in formal classroom training with levels of training comparable to that at an engineering college. (This education is carried on at a production training center that gives formal courses in physics, mathematics, chemistry, electrical and instrument theory, and equipment design and operation.) To qualify for promotion, a student must have six months of "hands on" plant experience and an additional eight months as an assistant operator being trained in sophisticated computerized devices that simulate control problems. Trainees are tested on more than one hundred simulated "events" that might possibly occur at an operating plant, such as malfunctioning relief valves, an accident indicating some measure of loss of the cooling water in the nuclear assembly, and the failure of a variety of backup safety systems or components.

I found it especially reassuring that the purpose of the extensive training through simulation of emergency events was to enable the operator not only to respond with corrective action but to acquire an understanding of the fundamental "behavior" of the entire system of the plant. The operator would then be able to assist in judging and correcting a broad range of apparently abnormal events, to know what to do when they flash on the simulator, and, whenever necessary, to respond with knowledge and confidence.

The Three Mile Island accident and the worldwide public attention it received produced the usual shallow rhetoric of public figures and commentators—"it must never happen again," and so on. However, the TVA board took immediate and substantive action, just as it had taken corrective action at once after the insulation fire at the Browns Ferry plant. One instance of that action was the decision to expand and strengthen the program for the selection of operators and their training in the "miniature university." The emphasis on stiff standards of selection and training has already led to emulation by many entities throughout the country. The elaborate TVA simulator facility has also led other atomic electricity enterprises to install simulators for the first time.

I cannot overemphasize the importance of careful operator

selection and training. During the many years of my professional and public responsibilities for technical enterprises, in the United States and abroad, I have emphasized one constant theme of people—the people who make technology work, the people who shape it (or fail to shape it) to meet human objectives and correct human error. I have seen machines get bigger and more complex, yet human beings still come in the same sizes and are still endowed with the same basic strengths, emotions, and limitations.

I know from experience the standards for the selection of operators of large, sophisticated chemical plants, such as those at Muscle Shoals, Alabama, for which I was once responsible as TVA chairman. I also knew well the supervisors and operators of many TVA hydroelectric plants and large coal-fired electric plants, as well as the supervisors and operators at TVA electricity distribution system centers. In later years, in my work in countries around the world, I kept a close watch on the selection and training of young men and women for similar technical jobs—in the Cauca Valley program of Colombia, modeled after TVA, for example, and in Iran, at one of the world's largest hydroelectric dams, a project in the building and functioning of which I had a large measure of responsibility.

By and large the selection and training standards I have observed in this country and abroad have been high and the results excellent—for the conventional industrial plant or utility system.

But I am convinced that the criteria for operator training with which over the years I have become familiar are not adequate in the case of atomic energy. The existing system of light water reactor installations requires much higher and more exacting standards than are now generally in use. Such standards can and must be developed; TVA has already taken a strong initiative in this direction. Utility managers and regulators the country over will follow this lead as they realize fully just how different is the very nature of atomic power from conventional energy plants, and therefore how much more time and effort must be taken to provide safeguards for plant workers and for the public.

A modern technical or industrial enterprise, whether a private corporation or a public bureaucracy, is by no means at all times a pattern of democratic methods; often it is dogmatic or autocratic. When the boss makes a mistake, he refuses to acknowledge it; he closes his mind, asserts his authority—and makes the same mistake again. In a democratic enterprise, on the other hand, there is the salutary custom of not only admitting one's mistakes but trying to learn from them. Sometimes, when things go wrong, the public has to give an indignant push in order to get corrective action taken. But whatever brings it about, American institutions generally show enough flexibility to make that change when they have to. The progress made in reducing industrial air and water pollution is recent evidence of this. The intense public attention given to the Three Mile Island accident should stimulate similar progress in the related field of atomic hazards.

It was reassuring to know that TVA did not wait to be pushed. It took a strong lead on the question at once. On June 1, 1979, barely a month after the Three Mile Island accident, the TVA board approved a declaration on atomic safety that is of historic importance. This document deserves to be recognized as the measure of how atomic safety should be judged and managed—a new national yardstick for public policy in this field. The fundamental point it makes is that "questions of nuclear safety [must] receive paramount consideration over and above cost and schedule requirements or operational needs of the system."

The safety charter, taken as a whole, is an extraordinary statement to come from an electric power producer (not just any producer, but the largest of all) with heavy responsibilities in precisely those areas—cost, schedule, and operating efficiency—

that it declares it will make secondary to safety. For timeliness, significance, and drama, I cannot recall anything quite like it in the history of the public utility industry. TVA, as a matter of declared policy, has put safety first in the nuclear power business —and fashioned its management structure to assure that the policy declaration will be implemented. For example, the board formed an "independent safety review staff outside the power, construction, and design organizations, which has direct access to the TVA board of directors," and created "a separate organization for nuclear organization . . . to concentrate on its unique problems."

It will not be easy. In the short run the additional cost could prove to be considerable. There may be pressures, from within the organization as well as from without, to relax the new standards, to cut corners for the sake of meeting schedules or for other reasons. But in the long run the whole country will benefit, for the TVA safety yardstick should become the measure for the entire atomic power industry. It can be an important example, too, for other industries that have concentrated for too long on markets and profits without demonstrating any great concern for safety and health.

Of equal significance, I think, is the spirit of candor and openness that pervades the TVA atomic safety program. The TVA board calls for the maximum possible internal discussion of nuclear issues and facts on the part of TVA employees, saying that "staff members with technical or professional views which differ with those adopted by management will be able to go directly to the independent Safety Review Staff and the Board with their concerns."

The TVA atomic safety document, which covers technical and operating details as well as broad policy issues, was drawn up by a special TVA task force that recommended immediate changes in existing practices on the basis of a study of Three Mile Island's deficiencies. This was a healthy exercise in self-criticism and self-correction. Altogether, I found the TVA performance in nuclear power hopeful and encouraging—and in sharp contrast to the complacent attitude toward the Three Mile Island accident on the part of many (but not all) of the other atomic power enterprises.*

*The Report of October 30, 1979, of the special presidential commission to investigate the Three Mile Island accident discusses the need for a change in the "attitude" of the atomic industry, and it uses the term *mindset* to describe that prevailing attitude.

# 8

# RADIOACTIVITY: A FACT OF LIFE
# —AND A HAZARD


On the heels of the Three Mile Island accident, a committee of
Congress was holding hearings on Capitol Hill on the hazards to
human health and life from radioactivity. From the tenor of the
testimony a listener might have concluded that any amount of
radioactivity, however small, was malign, and that by inference
this evil force was synonymous with atomic energy. One spectator
brought out a dosimeter, a familiar device that measures radioactivity. The instrument showed that the level of radioactivity in the
hearing room exceeded not only the level permissible under federal regulations in an atomic energy plant but also the level in the
environs of Three Mile Island following the accident.

What accounted for the reading on the dosimeter was the
natural and harmless radioactivity normally present in Washington, plus the radioactivity within the granite blocks of which the
building was constructed. If the hearing had been held in Denver,
at an altitude of five thousand feet, the increased cosmic rays at
that elevation would have produced a much higher reading than
that in Washington or at Three Mile Island.

The incident of the dosimeter in the hearing room is a homely
illustration of what is too often ignored: that radioactivity, or
radiation, is part of our normal daily life. Radiation is a fact of
nature, like the oxygen in the air we breathe, and like oxygen, it

can be harmful as well as beneficial to human health. But unlike oxygen and other nonvisible forces of nature, such as gravity, radioactivity has come to be widely portrayed in highly emotional terms as being malignant in and of itself, *at any level.*

This simply is not so. Massive radiation is hazardous, of course, but properly used, it can be beneficial, and it has been proven to be so over many years in the cases of diagnostic X-rays and therapeutic irradiation.

Since the beginning of time people have died by drowning and by fire. We have learned not to blame water and fire for these misfortunes, just as we do not blame gravity when an airplane crashes. Radioactivity, too, is a force of nature, yet we view it differently. We have known about fire and water, their benefits and their dangers, for centuries, but it is only recently that we have identified and given a name to the radioactivity that occurs naturally in the everyday world.

And it is more recently still that we have learned how to *create* radioactivity—in enormous amounts. The magnitude of this man-induced radioactivity is almost beyond comprehension. For many years radiation was thought of in terms of the familiar chest X-ray or a physician's use of radium for treatment. A gram of radium, derived from natural substances, is called a curie (after the Curies, the French scientists who "discovered" radium). For years the going price of a single gram of radium was $50,000, a measure of its scarceness. A curie is also a measure of radioactivity. Thus the reactor of a sizeable operating atomic energy plant, such as the one at Three Mile Island, creates fifteen billion curies of *radiation.*

It may take some time for the public to reach an adequate understanding of radiation, to know what is normal, harmless, even helpful, and what is hazardous. The jump from one gram of radioactive material, one curie, to fifteen billion curies of radiation is, after all, a sensational fact. And it is just one of the staggering facts about nuclear power—the force of man-induced radiation—which have perhaps inevitably given a science-fiction cast to the public view of the subject. But this sensationalism must be overcome if we as a people are to distinguish between what is harmful radiation and what is not, in order to provide a rational basis for determining public policy on atomic energy. That radiation can be controlled, in spite of the risk and the danger, is illustrated by the fact that for twenty-five years it has in fact been controlled in atomic reactors, with a few mishaps but without catastrophic loss of life or human injury.

Under normal operating conditions an atomic energy plant engaged in the production of electricity is as safe—safer, in my own experience—for the operators and those living in the imme-diate area than most industrial plants, and certainly much safer for health than coal mines and many chemical operations. It is safer than riding in automobiles on our highways, their record of tens of thousands of casualties every year a "catastrophe" that we lament but accept. The standards of safety for those working in nuclear plants are monitored constantly, and the results are far superior to those in most industries.

I speak of normal operations, when a plant is well designed and the operators are well trained; it is the abnormal condition that justifies the higher and more severe standards of safety. While after two decades the number of such abnormal conditions has been remarkably small, the goal remains no abnormalities what-ever.

Even in the very infancy of atomic energy, enough was known about radiation that its possible adverse effects on human health were treated with caution. In the 1940s, during my frequent official visits to the vast Oak Ridge centers or to the dramatic mesa in the mountains of New Mexico where atomic weapons were being fabricated, I was required, like every other worker or visitor, to don protective equipment and to wear a radiation-recording badge. As a part of our normal operations, we at the AEC at the very beginning set up a separate division that concentrated on safeguarding staff and employees from overexposure to radiation. As the years of experience and knowledge accumulated, the early precautions were not relaxed; on the contrary, the maximum time during the course of a month or a year that an atomic energy employee was permitted to be "exposed" to the radioactivity in these installations has been steadily reduced. The AEC's successor, the Nuclear Regulatory Commission, set five rems (a unit of human radiation exposure) as the maximum allowable exposure for any plant worker during one year; recently this was reduced by the TVA to four rems in one year for its own nuclear operations.

Radiation is not new on the earth, but man-induced radiation is, and our knowledge of the effects of radiation is still imperfect. Scientists have differed on the subject of radiation's dangers, and they continue to differ as year by year our experience and information increase. (Unfortunately, the amount of misinformation—to say nothing of publicity-seeking sensationalism—seems to grow as well.)

Among atomic workers, who number many thousands, radiation-caused illness or injury has been rare. No other major industry has kept as carefully documented a record of the health of its working force.

The chief public worry about nuclear energy concerns radioactive pollution of the air and water outside the plants. In normal operations, only minimal radioactivity escapes into the surrounding atmosphere. There is no guesswork here; radioactivity in the air can be measured precisely, and NRC regulations generally require careful, regular monitoring in the environs of atomic plants for a range of several miles, as I had occasion to observe firsthand several years ago in the vicinity of the huge, three million kilowatt TVA atomic plant at Browns Ferry, Alabama.

Because the potential hazards are serious, precautions are essential. In the history of the atomic utility industry there have been very few cases of the atmospheric emission of radiation, and none has presented a public health hazard, not even at Three Mile Island, where monitoring equipment showed that radiation pollution had been negligible.

The water pollution question must be considered in two parts. The electricity generating part of an atomic energy plant requires large amounts of water to be converted to steam, as does *any* generator of electricity based on steam-driven turbines. This water has no connection with the nuclear core of the atomic furnace, and therefore it contains no radioactivity. In that sense it is quite "clean." The use of water in the nonnuclear steam turbines, the electricity generating part of the plant, increases its heat somewhat, so its outflow does produce a moderate, intermittent rise in the temperature (but not the radioactivity) of the stream or river which supplies the water to the steam turbines. The steam that turns the turbines is routinely cooled, that is, condensed (converted again to water) in the massive cooling towers. This water is returned to the stream or lake, to be recycled into steam again when needed. With few exceptions, the minor rise in the temperature of the returned water is a quite manageable operating question of the kind dealt with routinely by all nonnuclear electric utilities that use steam turbines. The fact that the discharge of used water comes from an atomic plant involves no special risk for other downstream uses, such as public water supplies, industries, agriculture, or marine life.

The water used within the plants to cool the core is quite another matter. That water *is* radioactive, and even small quantities of it must be treated to remove all traces of radioactivity. One consequence of the mishandling of the reactor at Three Mile Island was that large quantities of coolant water poured into the containment building. The removal of the radioactivity from that water is necessary, and it is a major undertaking that will take months to accomplish, at great expense and always with some limited radiation risk to those doing the cleaning up. But there is no major technical obstacle to the operation; chemical decontamination, or filtering, is a well-known general industrial process that has a long history.

The need for this decontamination, however, must be considered one more of the serious inherent flaws of the water-cooled reactor design.

Quite apart from the operation of reactors, the atomic energy industry, for civil and military purposes here and in several foreign countries, employs tens of thousands of workers who in one way or another are exposed in their work to low-level radiation. The mining of uranium ore—itself a major industry in the United States, Australia, Russia, and elsewhere the world over—the transportation and handling of uranium ore, its processing, concentration and isotope separation ("enrichment") for peaceful civilian and military products, and the use of radioisotopes in hospitals and research centers—all these activities expose those who handle the materials to radiation. This kind of radiation is called low-level radiation; it is to be sharply distinguished from the intense, high-level radiation that occurs in a reactor.

There is a wide range of difference between scientific opinion and highly vocal nonscientific opinion regarding the effects of exposure to low-level radiation. The most extreme opinion maintains that *any* radiation, however "low," is a danger to human health that has caused cancer, leukemia, genetic defects, and other tragic consequences. At the other end of the spectrum of opinion is the view that the use of low-level radiation, in medical diagnosis or therapy, for example, while it does require knowledge and care, has not been harmful and its benefits far outweigh the potential risks involved. Low-level radiation is probably less "dangerous" than the emissions from burning coal.

Dealing with the fantastically high radiation intensities that are produced in the core of an atomic energy plant, or with radioactive spent fuel (the "wastes"), or with the handling and fabrication of plutonium for weapons, the risks are very substantial. It is my view that the present, almost universal method of utilizing this radiation to produce heat for the generation of electricity involves quite unacceptable risks for a long-term source of energy. I maintain that we must do more than try to patch up the defects; we must set about at once to find a method that has few or none of the present risks. And until that new method is found and tested, we must continue to work hard to reduce the present risks.

Risks there are, even with low-level radiation. But neither the country nor the world is well served by those who in the cause of personal notoriety seem to take relish in spreading fears that would induce a kind of radiation hypochondria and panic.

# 9

# WHERE TO LOCATE A NUCLEAR PLANT: THE ISSUES OF SITING

The location of nuclear plants—"siting," in the jargon of the trade—is at the heart of the worldwide public apprehension about atomic energy. When there is uncertainty and distrust and fear about nuclear radiation hazards, nobody wants an atomic plant in his back yard. Under those circumstances—the circumstances in which we find ourselves today, after more than twenty years of building nuclear plants that use the light water technology—no site is really a good site.

The best answer to the question of where to site such a plant is to develop a better, safer, more dependable way of producing electricity from nuclear fission.

But until we find that better way, we are dependent on the existing plants as a badly needed component in our electricity system. And we must decide how and to what degree we can go ahead with the more than ninety plants that are now in various stages of completion.

Many existing plants are located quite far from large centers of population, thirty or forty miles from cities of any size, in some cases even farther. The Florida Crystal River plant, for example, is fifty miles from Gainesville; the Arkansas Power & Light Co. plant, Arkansas One, is fifty-six miles from Hot Springs; the Edwin Hatch plant in Georgia is eighty-seven miles from the nearest city. Some plants, on the other hand, are close to population centers. The Three Mile Island plant is only eleven miles from Harrisburg, Pennsylvania, and the same distance separates Cedar Rapids, Iowa, from the Duane Arnold atomic facility. The Commonwealth Edison Zion plant is just seven miles from Waukegan, Illinois (and forty-one miles from Chicago); the Fort Calhoun plant in Nebraska is nineteen miles from Omaha. Con Edison's plant at Indian Point is thirty-six miles from the center of New York City.

The Nuclear Regulatory Commission has the responsibility and the authority for the continuous review of design safety features in the light of new technical information. (The studies now under way of what went wrong at Three Mile Island are the most obvious example of this.) It can order plants closed for safety changes, and it has done so in the past. The Three Mile Island accident and the subsequent public outcry have created a demand for a safety-first attitude that should encourage the NRC to exercise its watchdog function with great vigilance. It has already spurred utilities to strengthen their concern and actions for safety (as witness the most recent declarations and programs of TVA).

Still there are hazards. There are the risks of accident. The atomic power industry has an excellent safety record, and malfunctions have been few; the accidental release of radioactivity has happened in only a handful of instances and to a minor degree. But it has happened, and it may happen again. What has *not* happened—and what is statistically a far more remote danger to the ordinary citizen than being struck by lightning—is the heavy release of radioactive contamination outside a plant, over an area of dozens of square miles, as could happen in the event of an unstoppable burning out of a dying but still fiercely radioactive reactor core.

Of course people worry when they hear that such a thing is

possible. The motion picture *The China Syndrome* has added to the long list of waking nightmares with which a heedless and uncontrolled technology could conceivably confront us—melting ice caps caused by the carbon dioxide from the burning of coal, great lakes so polluted as to be lifeless, carcinogenic foods. In this dread catalog fact mixes with lurid conjecture, the known with the suspected. An atomic energy plant breakdown is a possibility, certainly; so is a hurricane's striking directly at some large city— New Orleans, say—or a flood of major proportions inundating parts of St. Louis or Memphis. There is no shortage of man-made or natural disaster possibilities to feed fearful speculation.

Yet the question remains, what should we do about our atomic plants?

The only absolutely safe course would be to close them all down.

In doing this, however, we would trade the remote potential of nuclear accident for the immediate reality of serious and crippling economic damage, vast human dislocation, and severe unemployment. Anyone who proposes turning off the atomic switch had better first figure out what to do with the hundreds of thousands of Americans whose jobs are dependent on the electric energy production that would be discontinued, the homes left without electricity, the streets darkened.

A more reasonable program (pending the development of a "clean" atom) would be to keep a vigilant watch on existing plants and those being completed, with particular emphasis on facilities that, like the Three Mile Island plant, are located near sizeable populations. TVA's lead should be followed in rigorous and creative operator training and in the redesign for greater safety of the various parts of the physical plant. The NRC should not hesitate to shut down either temporarily or permanently any facility whose operation cannot offer adequate safety guarantees.

Very well, the reader may say, but what *is* a safe site? How far from a large center of population should an atomic plant be located? No one can offer a hard and fast rule. If a plant is unsafe, it should be closed down whether it is ten miles or one hundred miles from the nearest city. Certainly no future plant should be closer to a populous area than is necessary for operating purposes, just as no plant should be built on an earthquake fault, in the flood plain of a large river, or in a coastal area that is regularly visited by hurricanes.

Siting is only one safety consideration, and perhaps not the most important one. I would feel safer ten miles from a well-designed, well-operated plant than I would sixty miles from a poorly designed and badly run plant.

Balancing the risks is a familiar process, one that we are accustomed to in this society, something we have been doing for many years. We accept without much protest tens of thousands of traffic deaths a year, and we don't close our highways. We put up with cigarette smoking despite the many thousands of deaths and shortened lives it has caused and is causing. People in California continue to live their lives atop an earthquake fault that produces an occasional minor tremor, despite warnings of major quakes to come. In the context of the risks we tolerate in our daily lives, some rational and others senseless, the acceptance of a measured and perhaps temporary risk in nuclear power does not seem to be an outrageous policy.

Where to locate a plant in order to minimize the risk and annoyance to the public is not peculiarly a nuclear energy problem, of course. In recent years the public and governmental agencies have become more and more sensitive to the location of all kinds of industrial operations, seeking to minimize the ill effects on the environment of smoke, poisons, and other effluents, and even noise and ugliness. There have been many grave instances of nonatomic pollution on a large scale in recent years—in the asbestos and herbicide industries, to cite two examples. The 1976 herbicide plant breakdown north of Milan, Italy, has seriously affected many hundreds of families, and the true scale of this disaster may not be known for a generation. Oddly enough, the public protest in the wake of this event was minuscule compared to the outcry against a proposed atomic plant at a location fifty miles north of Rome. Atomic energy has been invested with an air of black magic that makes it overshadow all else in people's minds, regardless of the facts.

Why does atomic energy have this power to darken the imagination? The answer, in part, is that the public is now fully aware of the hazards of the present early-stage nuclear technology. When those hazards are eliminated, people's fears will subside.

That is not the whole story, however. Some of nuclear energy's sinister reputation arises, no doubt illogically, from its historical origins. In the beginning there was The Bomb . . . and an official attitude of authoritarianism that was only partly justified by the requirements of security. The makers and shakers of the military atom ruled by secret decree and fiat, creating an elitist atmosphere that was to plague the civilian development of atomic energy for a generation. The experts knew all the answers and acted for the good of the public, whether the public liked it or not (or even knew about it). If the authorities in their wisdom determined that an atomic processing center or an experimental laboratory should be built in a certain location, there it was built.

The issue of siting an atomic facility brought about my first brush with the atomic age—the pre-atomic age, I should say, because at the time neither I nor anybody else, apart from a handful of scientists, was fully informed about a project for making the world's most destructive weapon out of something called nuclear fission.

The year was 1942, and I was TVA chairman. One day I was visited in my Knoxville office by one of the world's greatest physicists, Arthur H. Compton. To my utter dismay, Dr. Compton told me that the War Department had decided that it needed a large tract of land in the vicinity of nearby Clinton. He made it plain that I was not to inquire what the land was to be used for, that all farming must cease immediately, and that the farm families who owned the land must be moved.

I protested that the area was part of the Agricultural Test Demonstration Project; that the farmers had only recently been required to leave their ancestral land, which had then been flooded by the building of the Norris Dam on the Clinch River; that they had formed themselves into a community organization and were among the most progressive and successful small farmers in this part of the state; and that I simply could not tell them that they must once more give up their farms and community to the federal government. I told Professor Compton that there were other large tracts of land not so well developed where there would be no such disruption of people's lives as in the Wheaton community, and on a map I pointed out an area in western Kentucky.

No, said the country's leading physicist, the rolling lay of the land and the isolation of the Wheaton tract was just what was needed. In a few days the War Department would seek and receive a federal court order "condemning" all the land in a single transaction. I explained that this was an arbitrary action and not the way in which TVA acquired land; if I were to help explain this drastic action to the farmers and to the TVA staff, I would have to know the purpose. Dr. Compton told me politely but firmly that the entire undertaking was top secret and that I was to ask no questions.

Of course the Army got what it insisted on having. The tract was taken, the farm families were moved away, and the atomic industrial city of Oak Ridge was built. It was a bad precedent. That particular site was not essential; another, involving far less disruption in people's lives, would have served as well, but arbitrary bureaucracy, made doubly powerful by military secrecy, had its way.

It is ironic to recall that the secrecy, well intentioned as it was in wartime, did not prove effective. Secrecy rarely does. In a very short time, anyone who had any scientific grounding at all knew pretty well what was going on in the "mystery plant" at Oak Ridge and why so many leading atomic physicists were visiting there, even though their travels were supposedly hush-hush and they used transparently phony names.

Secrecy, with all its paraphernalia and ritual, has a way of holding on, leechlike, long after the original reason for it has ceased to be valid. So it was with this first atomic site. When Oak Ridge was built, it was a wartime secret. The whole place— houses, schools, and all, not just the atomic works—was surrounded by high fences, and guards at the gates turned back any visitor who lacked an official pass. After the war the Oak Ridge "secret" was out, of course, and there was no longer any reason why the town should be shut off from the rest of the world. I was AEC chairman then, and I agreed with the AEC city manager's recommendation that the town site fences come down.

There followed an astonishing phenomenon. A considerable part of the town's population wanted the fences! These citizens had no desire to surrender what they deemed a mark of superiority, something that set them off as special, separate, a privileged caste. They defended their fences so vociferously that I had to drop more important business and hurry to Tennessee to face a stormy protest meeting. (Our decision stuck, though; the fences came down.)

Elitism still crops up in atomic energy. Today there are serious

proposals to site clusters of nuclear plants at remote desert loca-
tions, hundreds of miles from any city, and to staff them with a
special cadre of dedicated personnel. This notion of putting the
hazardous atom far out in the middle of nowhere has a surface
attractiveness, but as a veteran utility operator, I can assure my
readers that however appealing it may be to theoreticians, it is not
a workable concept. A power supply so far from a service area, and
so totally outside the functional structure of this country's economic and electrical system, would be next to useless. Electricity
supply is a fundamental; it is integral to economic and community
development and to the livelihood and aspirations of thousands
of communities and neighborhoods. The present-day physical and
corporate remoteness of the electricity supply from the people
served accounts for not a little of the electricity industry's public
relations ailments.

Even if practical operating considerations and community
relations did not stand in the way of the cluster concept, I would
feel uneasy about such an arrangement. Nuclear power is too
important to be served up by an invisible cadre of experts.
Frankly, I would feel better about a plant that was not totally
removed from the vigilance of the public eye, whose operators
were not elite spirits but ordinary, intelligent, conscientious men
and women living in the communities served by their efforts.

The recently heard proposals for building atomic energy plants
deep underground, as a safety measure, display the same insen-
sitivity of the technocrats to the strong feelings of communities
and consumers about being "close" to their source of electricity.
It is this same concern that explains the cordial public acceptance
of small hydropower dams on local streams.

We have had enough mystery and secrecy and arbitrariness in
atomic energy. These things have no place in today's debate over
nuclear power, where all the issues, all the facts, and all the risks

should be talked about fully and openly and responsibly. Without that kind of candor and openmindedness, we will never rid ourselves of distortions and misunderstandings. One side will continue to minimize atomic hazards, and the other side will continue to exaggerate them; the public, caught in the middle, will be left in confusion and fear—the worst possible mood in which to determine a reasonable course of action.

# 10

# HOW TO HANDLE NUCLEAR WASTES

*The 500,000-year Fantasy*

An atomic reactor is potentially dangerous not only when it is in full operation. From time to time, usually once a year, it is temporarily shut down so that its old fuel rod assemblies can be removed and new ones put in their place. The old assembly is worn out only in a productive sense; it is still furiously radioactive, and its elements may be hazardous to human beings for hundreds of years.

The temporary storage of this material and its safe and permanent disposal has become perhaps the single most controversial subject in the nuclear debate. Many highly specialized scientists and engineers have developed a number of technical responses to the problem but few positive "answers." The whole area is fraught with technical and potential uncertainties. Atomic waste, after all, is something unique in science; it involves circumstances that have never before been precisely confronted.

Over the past three years I have studied the now foot-high stack of technical reports on this subject, the products of special government committees. (Such study groups usually end by recommending still another study.) With almost no exceptions, the indicated solutions call for deep burial of the wastes, encased in glass or ceramics or steel, in stable mineral deposits such as granite or salt. No single method has received the unequivocal endorsement of the scores of noted geologists, earth scientists, and public officials who have examined the problem. On the other hand, the general assumption is that we must settle on a permanent solution and put it into effect as soon as possible. According to this widespread feeling, we cannot permit atomic wastes to accumulate any longer. Everything depends on discovering a way of getting rid of this dangerous material. If we fail to find a safe solution, nuclear power simply "has no future," as one distressed official put it. In short, nuclear power will be finished and the plants will have to be shut down.

Or will they? Not everyone agrees. I certainly do not.

The range of expert opinion and public and political rhetoric tells me how unprepared we are at present for making decisions on a "permanent" disposal plan. I will cite two examples. Dr. Alvin Weinberg, one of the fathers of atomic energy, dismissed recently the widely publicized fears on this score as unfounded. Major problems of safety surely exist, he indicated, but this is not one of them. Quantities of such material have in years past been buried in a deep hole at the Oak Ridge atomic reservation, with no evil aftereffect. The total quantities involved are very small. Processed into a solid form and buried, the noxious stuff in a thousand years will by the predictable "decay" of the radioactivity be innocuous to man, no more radioactive than the natural uranium in the earth, from which it was originally mined. Dr. Weinberg, not unaware of the public clamor, dismissed the problem as little more than a "nuisance."

On the other side, a spokeswoman for a highly influential and respected environmental organization, the Sierra Club, has written that the test of a permanent disposal plan is whether the place where the material is buried will be able to withstand the disturbance to the earth of a war a half million years in the future, the disruptions of the "geologic folding of the earth," or ice sheets moving down from the Arctic Circle.

Neither approach is persuasive. Despite the limited amounts of material involved, the problem of atomic wastes seems to me to be far more than a "nuisance" that can easily be disposed of. Nor does the 500,000-year argument make much sense to me; it is so overdrawn and science-fictionish that it does not seem intended to be taken seriously—a debater's trick, not an attempt to deal realistically with the problem.

Yet a clear policy on atomic wastes is certainly called for. Public agitation has been mounting rapidly in recent years. In the absence of a rational program, irrational measures may be adopted. Public bodies are being bombarded with demands that they do something. Not only Congress and federal agencies, but even state legislatures and several governors have taken the initiative in responding to this public concern. Some have sought to declare a moratorium on new plants; others have aimed at forbidding the use of highways for the transport of waste fuel from plants to storage sites. In the continued absence of a national program, a multiplicity of state and local regulations will make arriving at a workable and rational program extremely difficult.

The state and local actions, while understandable as expressions of anxiety about this disturbing hazard, do not really touch the heart of the problem. Closing all the nuclear energy plants would not resolve the waste issue; fuel wastes would continue to accumulate. For the bulk of the waste comes from military sources, created in the course of producing nuclear bombs. This was how atomic wastes began historically, as bomb material by-products, and since there is—God help us—no end in sight to this fearful industry, the wastes will continue to pile up.

We should not allow ourselves to be deluded on this point.

Switching off the nuclear plants would not meet the issue squarely. Wastes we have, and we will have more. We cannot abolish them; we must find some way of coming to terms with them here and now.

## More Bombs, More Wastes

The disposal of atomic waste material is now a hot and lively issue, but there was a time when it seemed to be almost irrelevant.

Not long after the new civilian Atomic Energy Commission was created in 1947, I and my fellow commissioners set out on a memorable tour of inspection of the atomic facilities built during World War II for the Army's Manhattan Project. This included the world's first plutonium production "pile" at Hanford, in the arid reaches of south central Washington State.

What we saw was in fact an atomic energy plant, for its reactor core was seething with the tremendous heat of radioactivity. But this heat was not intended for the production of electricity. It produced plutonium. The bomb that on August 9, 1945, destroyed the Japanese city of Nagasaki, killing thousands of men, women, and children, was made of little more than two kilograms of plutonium that came from Hanford's ugly stack of graphite blocks, the "pile."

I had first visited Hanford in 1946, in a far different capacity. Then I was chairman of a State Department board of consultants on international atomic energy control. If the American Plan for International Control of Atomic Energy, based on our report, had been accepted by the Soviet Union, as it had been by virtually all other nations, that might have closed down the Hanford "pile" and put an end to its grim function. Atomic energy would then have been developed for peaceful uses only, under a system of United Nations control and management.

But in 1947 we of the new AEC had little time for thoughts of the "peaceful" atom. Russia, which at the time had not yet learned how to build atomic weapons, had rejected the American UN control plan and was busy trying to build atomic bombs of its own. And we had just reported to President Truman that America's supposed shield of A-bombs simply did not exist. Under those circumstances, our sole preoccupation was to step up plutonium production for the fabrication of atomic weapons. And it was not long before we had more than forty thousand construction workers building a new Hanford plant under the management of Du Pont. Disposal of the spent fuel assemblies from the plant reactors was considered a secondary matter, and the question of producing electricity from such reactors still lay in the future.

(An interesting sidelight on the question was the outspoken opposition to nuclear power on the part of the two men most responsible for the successful production of plutonium at the wartime Hanford plant, Vannevar Bush, President Roosevelt's scientific counselor, and James B. Conant of Harvard. Both men feared the radiation hazard from atomic wastes. The usually serene Conant was passionate on the subject, and he once exhorted me to head a "crusade" against nuclear power development. Not all scientists agreed. The great E. O. Lawrence had talked to me enthusiastically about nuclear power prospects in 1946, saying we had the technical means to do the job in a year or less. As it turned out, the first commercial reactor went into operation in 1957.)

Under the Army's Manhattan District management, the poisonous radioactive wastes from the giant Hanford plutonium plant had been poured into stainless steel tanks and buried in trenches on the Hanford reservation, where they remain today. We of the new AEC were told—accurately, it turned out—that the tanks should be able to resist corrosion for "a very long time."

We asked, How long? The answer: Oh, certainly twenty years or so.

Well, twenty years then seemed a long time in the future, and the matter of no great importance when so many other tasks were pressing urgently for action. For years the AEC continued to put its toxic atomic wastes into heavy metal casks and to bury them at Hanford. This proved to be but a short-term solution—and a poor one, for in recent years some of the earliest tanks have started leaking waste into the sands of the site.

Reflecting now on the present fierce intensity of public feeling about atomic waste disposal, it is a little difficult to recapture the wholly different mood of that earlier time, the late 1940s, when atomic electricity did not exist, and when those of us responsibly concerned were haunted by the specter of our slowly growing but still quite scantly furnished A-bomb arsenal. This was our great worry, and of course how many bombs we had (and didn't have) was highly secret information. I remember in 1949 making a formal report on this subject to President Truman in the presidential office. Following our elaborate security practices, I had avoided putting down on paper the true figures relating to our bomb stockpile. The figures on two tiny separate slips of paper I carried were deliberately incorrect, a perhaps theatrical bit of security precaution. As I prepared to give the President the correct numbers orally, Defense Secretary Louis Johnson outdid me in the matter of supercaution. There we were with the chief executive of the United States, in the absolute center of power in the non-Communist world, surrounded by the densest possible thicket of antispy security devices, and Louis Johnson hastened to warn me, "Don't read the figures out loud." (As I recall, I did in fact lower my voice.)

That was the spirit of those times.

We were worried then about letting the Russians know what

a pitiful handful of bombs we had. And now? Just the other afternoon I heard public testimony on the SALT II treaty proposal, in which it was disclosed almost casually that we now had ten thousand nuclear bombs (as did the Russians). Not only that; under the proposed treaty, we and they could build up this ghastly figure to sixteen thousand bombs each.

The amount of long-lasting radioactive wastes from this fantastic present and prospective military production makes our early waste disposal problem seem small indeed.

## A Short-term Solution

Removing from an atomic reactor a fiercely radioactive and deadly fuel assembly (typically twelve feet long) is an industrial operation utterly unlike any other that I have ever witnessed; indeed, it is unique among industrial procedures of the kind found in steel mills or chemical plants. Special clawlike devices, designed with an awareness of the dangers of high-level radiation, are used by trained technicians to grasp the fuel assembly rods and withdraw them, the operator's hands, remaining in protective gloves, thrust through a sealed, radiation-proof barrier. This procedure calls for skills and reflexes of a high order. Yet over the years since I first witnessed the procedure, a considerable number of people have mastered it. Except for one fatality at the Los Alamos fabrication operation—a man who deliberately ignored instructions—hundreds of thousands of hours of this handling of waste spent fuel have gone by with no other injury to operators and no escape of radiation into the world outside the plant.

This operation is followed by one that appears even more eerie to the observer: the claws deposit the still fiercely radioactive fuel rods in a nine-foot-deep rectangular pool of water, which becomes suffused by an unearthly bluish-purple glow.

When I first witnessed this remarkable display of technical skill, I was impressed by one particular fact—that those radioactive spent fuel assemblies, plunged in deep water, were thus rendered harmless and would *remain* harmless as long as they were kept there.

These assemblies contain valuable material—uranium, plutonium, and useful isotopes. Chemists have developed methods for reprocessing these residues, converting them to liquids and solids that can be used as atomic fuel for other reactors in the production of electricity, for plutonium weapons, or for medical purposes. The extraordinarily successful record of safety in the industrial handling of these materials and the training of many hundreds of technicians have given assurance that a reprocessing industry could be established, although the economics of such reprocessing for private investment is still marginal. (Two privately owned plants for such reprocessing, one in New York State and another at Barnwell, South Carolina, were built. They failed to be profitable enough to be kept in service.)

There is, however, a grave *political* impediment to the "mining" of valuable elements from radioactive spent fuel assemblies. The fact that the primary bomb material, plutonium, can be recovered from these wastes means that any nation possessing nuclear power reactors that use our present technology is automatically given a potential nuclear weapons capability. This threat gives a double edge to the question of waste hazards—that is, the wastes are dangerous not only in themselves but for the plutonium that can be extracted from them for weapons use.

Many methods of temporary and permanent storage of atomic wastes have been tried or proposed. Some of them—encasement in ceramic material, for example—seem promising. Other and better methods are certain to be developed in future years by the physicists, chemists, and nuclear engineers who are working on this vital problem. The story of science and technology has in fact this constant theme: if today's way of doing it isn't satisfactory, just wait; there'll be a better one tomorrow.

But can we wait? Do we have a safe way of storing atomic wastes on a temporary basis now, while we wait for some future permanent answer?

I am convinced that we do. We not only have such a safe means of temporary storage, we are using it today. That the public is simply not aware of this fact accounts for much of the present worry about radioactive waste.

The spent fuel assemblies from all nuclear reactors, military as well as civilian, are placed for temporary safekeeping in deep pools of water right at the plant site, as soon as they are withdrawn from the reactor core.

In the past these spent fuel assemblies were removed periodically from pool storage and transported to the two giant reprocessing centers in New York State and South Carolina. The real or imagined hazards of transporting this material on the highways or railroads touched off widespread protests. This, and the economic and political impediments to reprocessing just described, virtually put an end to waste transportation—and created a situation of stalemate.

The spent fuel assemblies have thus been kept in the pools of water at the respective plant sites simply because there was no practical alternative.

I say that this situation is a quite acceptable and safe *temporary* solution of the waste problem.

Some of these storage pools have been in safe use for many years. No reason has been advanced by anyone with operating experience to explain why, with proper monitoring and maintenance, they could not continue to retain safely the spent fuel assemblies for at least another score of years. The water in the pools does not become dangerously radioactive because of the fuel rods; simple housekeeping maintenance and security measures have been all that is needed to assure safety to human beings.

During the period that the rods remain in these pools the technical, managerial, and governmental issues of waste disposal could be fully considered in the light of technical advances that will be made in coming years. Thus *permanent* disposal operations may in time be undertaken with the care and foresight that the importance of this decision warrants. (Not the least prickly issues will be the definition of the respective roles of federal and local government and the determination of who must pay the costs.)

Ingenious engineering methods have been devised to allow existing pools to hold far greater amounts of spent fuel. Yet the present pool system will have to be enlarged. In most cases this will present no insuperable problem, for the majority of reactors are sited far from populated centers, where additional land can be acquired fairly readily.

The space requirements, in any event, are not great. I have received a reliable estimate that a 1,000-megawatt reactor produces in one year's operation only 3.1 cubic meters of solidified high-level radioactive waste. Space is not the decisive limiting factor once the material has been solidified and concentrated.

Keeping waste fuel in deep pools of water *at the reactor sites* for at least another twenty years has a number of advantages (in addition to the fact that the safety record of pool storage is a long and unblemished one).

The alternative—choosing some means of waste containment now and burying the material deep in the earth—would preclude a future decision to resume reprocessing. We would lose forever large and valuable amounts of uranium, a commodity that might be in short supply in another few decades. The pool solution, which keeps the wastes on the reactor sites, would also eliminate the controversial waste transportation issue. In addition, wastes stored in pools can be readily monitored from time to time by nuclear technicians, and this would not be possible if the wastes were sealed and buried.

I believe we should recognize the fact that we now have a safe and effective means of *temporary* waste storage that will buy the years of time we need in order to resolve the problem on an enduring basis.

# 11

# THE GROWING PAINS OF
# REGULATION

The sudden coming of the atomic age shook not only men's minds but also their institutions. This profound new force tested the strength of our society; its impact on the relations between nations was harsh and immediate. Almost overnight we and the rest of the world, just emerged from the most terrible war in history, were forced to come to terms with the overwhelming threat of a totally new dimension of power.

All of this was recognized from the very beginning. In the field of international relations the need for new institutions and the drastic modification of existing ones was most apparent. My life with the atom began in the autumn of 1945 and was concerned with this aspect of finding a way to control internationally what at that time was thought of almost entirely as a weapon. It was widely believed that the fate of the newborn United Nations rested on the outcome of this issue. If the UN member nations could not agree on a way to control the destructive new force, it was said, then the whole concept of an international community was doomed. In fact, as everyone knows, the comprehensive American atomic control plan was not accepted by the Soviet Union, and the future effectiveness of the United Nations was gravely compromised. In later years more modest efforts at international control were undertaken, with some success but within a very limited framework.

The question of international atomic weapons control is certainly related to American domestic atomic energy policies; however, I can do little more than touch on this issue in the present work. This chapter concerns the attempts of the United States government to regulate and control the development of atomic energy within its own borders.

A greater test of American institutions could hardly have been imagined. The controversy over how atomic energy is to be controlled in the public interest, how it is to be regulated, how its governance is to be divided between the President and the Congress—these questions have been debated and legislated about from the very beginning. It is a measure of the difficulty of regulating such a new and revolutionary source of energy that this debate is still going on, more than thirty years later.

The public and particularly the political leadership of the country quickly recognized the sensitivity as well as the complexity of the atomic energy issues. Some of the most strenuous and often most painful experiences of my life in public service illustrate how atomic energy was perceived to be a turning point in the relations between public authority—notably the Congress—and the scientific, industrial, and military establishments.

The first major controversy concerned whether this scientific leap into a new world should continue to be managed by the military. The logic for continuing military control seemed obvious to many, including some of the leading atomic scientists. After all, the Army had administered the Manhattan District, which had produced the A-bomb, a fantastic achievement that had revolutionized warfare. Atomic energy was weaponry, and weapons, it was said, were the business of the military establishment.

This view was strongly opposed by many, including members of the scientific community and, most important, President Truman, who urged repeatedly the development of the peaceful potential of atomic energy as a counterweight to its dread destructiveness. Truman had sharp and positive opinions about keeping the military under civilian control (as he demonstrated in his famous dismissal of the insubordinate General Douglas MacArthur). The atomic issue was debated in the nation at large and in the Congress; the outcome was a victory for civilian control, embodied in the McMahon Act, which established the all-civilian Atomic Energy Commission.

An equally far-reaching debate was touched off by my nomination as the first AEC chairman. This was over the issue of the government's role in the control and development of the new source of energy, important to industry generally, and specifically to the very influential private electric utilities. To opponents of TVA—there were many, and they were well organized—the appointment of the chairman of the publicly owned TVA to head the new AEC was the signal for an acrimonious controversy. (There is irony in the fact that years later, almost immediately after leaving the AEC, I urged publicly the phasing out of the existing government monopoly of atomic energy, arguing that the technical development of the atom for electricity generation should become a function of private enterprise—which subsequently happened, to a large degree. My basic point was that the development of a new technology was traditionally a function of the private sector and that such development would be stunted if it remained exclusively a government function.)

The question of civilian versus military control was settled early; the question of government versus private development was tackled, if not resolved. There remained other questions, large among them the problem of how to manage this awesome new force—how to safeguard the public interest, how to administer the atom for the general welfare.

The new AEC, the initial administrative answer, was highly unconventional in terms of American public institutional practice. It was an agency created with an unprecedented range of authority and responsibility that covered both the development of peaceful atomic energy and the production of atomic weapons. The latter, by far the more important part of the job in everyone's eyes then, brought about an immediate need to work out effective relations with the military establishment. It was a most unusual situation. We of the AEC, civilians all and responsible directly to the civilian chief of state, were manufacturing weapons for military purposes and holding them in our custody. A good deal of my time in the AEC was spent wrestling with the day-to-day problems that this peculiar but unavoidable relationship presented.

(One of the most dramatic episodes I can remember was the confrontation before the President in 1948 between the top Defense Department officials and the AEC commissioners. The issue was whether the AEC should retain custody of what by then had become a substantial stockpile of atomic weapons or whether the military should take possession of them, as it vigorously proposed to do. Truman again stood fast for civilian control, and the AEC retained this vital and sensitive responsibility.*)

Legislative supervision and control of atomic energy matters presented many new, challenging, and often puzzling problems.

Congress created a special Joint Committee on Atomic Energy. Such a committee of both houses was not without precedent, but the powers of this particular committee were broader than those of earlier joint committees, thus matching the unprecedented character of the Atomic Energy Commission itself as well as the extraordinary nature of the problems posed by the underlying scientific discoveries.

---

*I described this dramatic confrontation at length in my journal for July 21, 1948. See *The Journals of David E. Lilienthal*, Volume II, *The Atomic Energy Years, 1945–1950*, New York: Harper & Row, 1964, pp. 388–392.

We of the first AEC were required by law to keep the congressional joint committee fully informed on atomic matters. Yet the statutory obligation to maintain secrecy with respect to certain vital matters was often at odds with this requirement, making it difficult at times for us—and for the members of Congress concerned—to carry out our obligations properly.

This contradiction was immediately evident in our relations with the House and Senate Appropriations Committees, whose approval of our budget requests was necessary. Our fund requirements were very large, for they included not only research and development costs but the expenses of the production of weapons material for the military. The law called for the strictest secrecy on weapons matters. Thus, as AEC chairman, I found myself appearing before these committees, unable legally to provide anything except the overall lump-sum figures we were requesting. To provide explanatory detail of the kind I had been accustomed to presenting to Congress during my TVA days would inevitably have disclosed the kind of sensitive, top secret information that the Atomic Energy Act declared could not be divulged even to a committee of the Congress. How could I provide the kind of information that would give an appropriations committee a fair opportunity to pass judgment on the amounts requested? The good sense of Appropriations Committee members helped to find a pragmatic case-by-case resolution of this problem of public control. But the problem remained: how can judgments be made if knowledge on which they must be based is unavailable?

The power to appropriate (or to refuse to provide) funds for public programs is at the heart of public control. Some members of the appropriations committees at times requested explicitly that they *not* be given vital information; their thought was that if there were a "leak," suspicion might rest upon them. In contrast, the chairman of the congressional joint committee was incensed when I refused to answer his questions about the number of atomic bombs we actually had—this information being a high order of top secret.

During my tenure as chairman of the AEC I was most troubled not so much by this kind of aggressiveness toward the AEC by members of the joint committee as by instances of triviality and ignorance. Pettiness in the exercise of the serious legislative duty of regulation and control is bad enough when routine activities of the government are concerned. At times it was devastating when it was exhibited toward so complex and technical an undertaking as atomic energy, particularly for the damage that was done to the prestige of government in the eyes of the talented scientists on whom this pioneering work depended.

As I write, the country continues to be faced with the inherently difficult problem of public control of a new and basic realm of science and technology. The point that needs stressing here is that public control and regulation of a technical and scientific enterprise demand that the regulators show the normal, fundamental courtesy toward those they regulate by informing themselves adequately about the subject involved.

One widely publicized instance will illustrate the point. In June 1949 the congressional joint committee, with great agitation, summoned me and my AEC colleagues into executive session shortly after the publication in New York and Washington newspapers of a sensational "exposé." The story was that "a bottle containing uranium" was missing from the AEC's Argonne National Laboratory outside Chicago; the facts were that the substance involved was not uranium but uranium oxide, and the quantity was a minuscule amount. But the committee was in an uproar, as though Russian spies had cleaned out our bomb lockers. This was too much for me. I told the committee, "Criticize the Commission if you wish, as much as possible, but don't induce hysteria in this country. This is not bomb material, and you should say so, and it is small in any terms. If the people of the country find that the Congress is rattled over a seventh of an ounce of uranium oxide, what kind of hysteria in the Congress can we expect when the day comes when Russia has a stockpile of atomic weapons?"

The committee proceeded to investigate this meaningless episode, wasting an enormous amount of everybody's time, until the oxide was finally located a few days later. Dr. Walter Zinn, head of the Argonne Laboratory, spent an entire morning testifying about whether the bottle was made of brown or clear glass and other such vital matters! I wrote in my journal at that time: "I have been more depressed about this job this morning than I have been for a long, long time. Chiefly, I suppose, because all this seems such a frittering of time and emotion and energy, such triviality in a world that is really faced with great and grave troubles. Such a waste of good men's minds."

After I left the AEC the joint committee began dealing directly with the commission, not in its assigned statutory watchdog capacity but in a managerial role. Far-reaching technical conclusions about the kind of reactor that should be financed, the kind of research program that should be pursued, and the whole range of technical and managerial problems (the responsibility for which was vested by law in the AEC) became instead a part of the everyday functioning of the joint committee. Some of the most important decisions of a highly technical character were taken from the AEC upon the "prompting" of the committee.

This gradual absorption of executive authority by representatives of the legislative branch had long-term implications for the course of the atomic energy program—and particularly for the Achilles' heel of atomic energy: safety.

In the early 1960s this hybrid legislative-executive management—the joint committee, the AEC, and their respective staffs —made a far-reaching and decisive commitment to the light water nuclear reactor technology, which appeared to offer a means of producing electricity that was cost-competitive with coal-generated power. For the next decade there was enthusiasm and large financial outlays by private manufacturers and electric utilities for the purchase of the light water reactor. There was almost equal enthusiasm for this type of reactor in foreign countries. Words of caution about safety had no dampening effect; orders poured in for the American reactors. Among the companies that benefited the most was the designer and manufacturer of the Three Mile Island reactor, Babcock & Wilcox.

It was at this time, in 1963, that I criticized publicly the bandwagon rush to the light water reactor before its safety had been proved. My criticism extended to the role undertaken by the AEC in those atomic boom years, which in my view did not protect the public interest. That is, the AEC (with the stimulus and partnership of the joint committee) had become the active and enthusiastic *promoter* of atomic power at the same time that it was supposed to be exercising the quite different function of *regulator*. Here was a clear conflict of interest; in government and in law, the roles of judge and advocate must be distinct.

The ultimate demise of the AEC—and of the joint committee —can doubtless be ascribed to various factors, but it seems probable to me that this unhealthy mixing of functions had a lot to do with the congressional action in 1977 that effectively terminated both bodies.

The successor and current administrative arrangement, the Nuclear Regulatory Commission, is a further experiment in safeguarding the public interest and the public safety in the field of atomic energy—an enterprise so largely public in nature and so broad in its consequences, and yet which for the most part is carried out by private industry—the electric utilities and the equipment manufacturers.

The Nuclear Regulatory Commission was created to take over the regulatory functions originally vested in the AEC; it was (in theory, at least) to have no role as a promoter of atomic energy. The research and production functions of the AEC were transferred to the newly created Department of Energy. The joint committee's legislative functions were taken over by other committees of the Congress.

The regulatory devices of the NRC are the comprehensive and detailed licensing of every stage of the design and construction of nuclear plants and the supervision of the plants after they are functioning. This new system received its severest test in the shock of the Three Mile Island accident.

In my opinion, an opinion shared by many, the NRC did not pass the test imposed by the Three Mile Island accident. But it is so easy to lay the blame for any and all shortcomings on a public agency or a utility. I find it sophomoric and a bit cheap to single out the obvious inadequacies of the NRC—its structure, its personnel—as being largely responsible for the near catastrophe. The proliferation of second-guessing and the rise of newborn "experts" with no practical experience either with atomic energy or with public regulation has increased the volume of criticism of the NRC without throwing much light on an intrinsically difficult problem, the public regulation of a new technology.

It is premature, it seems to me, to propose a radical or final solution such as nationalization, to the various and difficult problems of public regulation. The field is still quite new; it is complicated with technical and political considerations. It is not, however, premature for me to raise some fundamental issues and to express opinions, based on my regulatory and operating experience, concerning the regulation of this new industry.

The accident raises questions of public control that are of major importance, that go far beyond the issue of whether the commission should be run by a single administrator or by five commissioners, and it is an irresponsible reaction to the accident to brush aside fundamental questions that have been raised by the faults of the technology itself.

We must stop playing word games with the future of atomic energy and begin to discuss substantive issues. For example, is it sensible to impose a criterion of safety ahead of profits or costs when atomic plants are operated (as most of them are) by investor-owned, for-profit utilities; or when the operation is in the hands of a publicly owned enterprise in which public opinion and political pressure to hold down rates and costs may influence operating criteria for safety?

Many new questions have arisen about atomic energy, questions that, in my opinion, a change from private to public ownership—or the reverse—would not resolve. Whether atomic energy were publicly owned or the energy provided by profit-oriented utilities would not have affected results in the recent stormy years of widespread public opposition to atomic plants. How should a federal public regulator such as the NRC take into account and be responsive to the apparently strong public feeling that the people of a locality, a state, or a region affected by an atomic plant must have a determining voice in matters of safety involving a particular plant in such a location? Is it in the cards that a private, investor-owned utility come up to the demanding job of operating atomic plants under the rigorous and very expensive standards necessary for public safety?

My own view today is that proposed "solutions" such as nationalization of existing and new atomic energy plants would complicate the problem of safety and regulation rather than help solve it. This is a sophomoric and strictly political response to the basic technical engineering and administrative problem; it has very little to do with the *basic* problem: the suitability of the technology for producing power from the atom.

Our existing complex nuclear technology presents particularly onerous difficulties for sound regulation and efficient management. The engineering has become inordinately complex—far too complex. This complexity is not the mark of progress; quite the reverse, it is a sign of relatively primitive development. The early radios, for example, were marvels of ingenuity in their way, but they were huge and unwieldy compared to the devices we now have, which are both sophisticated and simple.

When the engineering of a hazardous technology is complex, the regulation required for safe operation tends to be complex as well. For there are thousands of items to be checked on instead of a few key ones, and regulation must be extended to the minute details of design and construction and even of daily operation. Consider an actual case: an NRC substaffer decides, with not a little arrogance, that in tightening a screw or bolt, three rather than two turns is correct and "safe"; thereupon a written "procedure" for screw tightening is provided, under the guise of federal "regulation." Regulators (who are no more competent than the operators) look over management's shoulder on every detail; in effect they become a second management. And anyone who has ever had executive responsibility knows that when you have two managers for the same job, you have no management.

Any examination of this key area of public interest, the regulation of nuclear energy, makes us look ahead all the more impatiently to the development of a better, less complex, less hazardous way of producing electricity from atomic fission.

Our present system is not only cumbersome in a scientific and engineering sense, it is a manager's and a regulator's nightmare.

# 12

# AN INDUSTRY IN ITS ADOLESCENCE

The impact of new discoveries on people's lives was a slow and gradual process in bygone times. The voyages of Columbus and Magellan and the invention of the telescope, the printing press, and the steam engine all had profound consequences for human society, but their full effects were measured out over many decades, even centuries.

The rate of change—the speed with which a basic discovery is exploited in practical ways—increased with the beginning of the industrial age, yet even then it was moderate by today's standards. More than forty years passed between the testing of prototype locomotives and the railroad construction boom of the 1840s. The widespread application of Edison's electrical inventions required more than a generation. Even the gasoline-powered automobile had a long childhood; invented in 1862, it was not turned out in appreciable numbers until the opening of this century, and only in 1913 did Henry Ford's celebrated assembly line begin mass production.

Since World War II the rate of change has gone from an easy canter to a mad gallop. Nowhere is this more evident than in the field of atomic energy. The first commercial prototype reactor went into operation in 1957; by 1962 atomic fever was raging and plants were being built all over the country.

We hear it said—and by well-informed and serious people, disillusioned with atomic energy—that the nuclear industry "has no future." However true (or premature) this judgment about the future may be, one must keep in mind the fact that atomic energy has had only a brief past. It is an adolescent industry. Its technology remains immature. How could it be otherwise in such a short span of time? Two decades are not enough for the full and responsible development of a major new industrial sector that calls for new knowledge (of metallurgy, for example), levels of workmanship, and manufacturing techniques of a higher degree than any previously known or practiced. The pace has been a dizzy one. There are areas in the United States and in several other industrial countries—France, West Germany, Britain, Japan—which now obtain a large portion of their electricity supply from a source that just a relatively few years ago did not exist.

In reaching a judgment about atomic energy today, we need to remember the high hopes that the beginnings of the peaceful atom produced. In 1945 a congressional declaration of policy confidently predicted that "tapping this new source of energy will cause profound changes in our present way of life"—which changes were expected to be for our common good.

At midnight on December 31, 1946, a document signed in my presence earlier that day by President Truman became effective. It established the new civilian Atomic Energy Commission, responsible for developing peaceful uses of the atom, including atomic power. On that same day General Leslie Groves, head of the wartime atomic bomb project, to whose duties I succeeded, expressed a commonly held view of the potential of the peaceful atom when he said to his former Manhattan District associates, "With regard to peaceful application, you have raised the curtain on vistas of a new world."

The prospect that humankind had at its disposal a "new

source of energy" unlimited in its possibilities fired the creative imagination and helped to ease the sense of guilt for the efforts of science having been put to so inhumane an end as the holocaust of Hiroshima. There was on all sides a great impatience for immediate results, an impatience greater than that which accompanied almost any other discovery in history.

Since that time very substantial progress has been made toward realizing the prospect of a "new world."

It is, however, my view and the view of some proponents of and participants in the atomic energy field that we tried to move too rapidly, that not enough time was taken to test reactor ideas and designs that were radically different from the light water design that quickly preempted the market—alternatives that might have provided us with greater safety and certainly far less complexity.

Today the light water reactor, with its dangerous side effects, is virtually the only nuclear generation facility available to the electric utility industry.

I know of no historical precedent for this anomaly: that to solve a major technical need there should have been almost universally adopted, virtually from the beginning, one and only one method, one technology. In contrast, the early years of the automobile industry, for example, saw literally dozens of companies competing among themselves with radically different designs. Cars were powered by steam and electricity, not just gasoline. Indeed, for years gasoline was in third place. Even after the gasoline engine became predominant, the others persisted. (Steam autos were made as late as 1926, and there have been periodic efforts to revive this method. A real rebirth of the electric automobile now seems imminent.)

In atomic energy several different technologies were in fact proposed and considered in the beginning. And they were soon put aside. Whether any of these would have proved to be better, no one can be sure. Apparently they were considered to be more costly to make and therefore less profitable to the equipment vendors. But might they have been safer? Who now can tell?

Among the many possibilities were a high-temperature, gas-cooled reactor; a natural uranium reactor; a molten salt reactor; several versions of so-called breeder reactors; and a heavy water reactor. These proposals offered different methods of heat transfer and moderating and controlling the nuclear fission reaction in the reactor core and of transferring the intense heat through steam to the turbines and electric generators. They should now be reexamined, especially in terms of safety, and given a thorough testing. New methods beyond these can and should be developed, particularly if the atomic industry can rid itself of its fixation with the basics of the light water reactor as being the absolute and ultimate answer, come what may.

I should say that new methods *will* be developed. That is the lesson of history. Scientific research and technological development may be stymied for a time, stumped by particular problems, but sooner or later the breakthrough comes, often in unorthodox ways and from unexpected directions. This process can be seriously delayed, however, by the cautious mentality prevalent among some sectors of the atomic establishment today.

The reliance on a single, virtually unchanging technology is contrary to everything we have learned about scientific and industrial development—indeed, about virtually any aspect of human endeavor.

What happened in the field of radio is one of many illustrations that come to mind. At one time radio was completely dependent on what was considered a marvel: the vacuum tube, a kind of electronic valve. The conventional approach to the improvement of the radio on the part of manufacturers was to make better vacuum tubes. But the Bell Laboratories, driven by its own need to develop a better answer for the communications field, eliminated the vacuum tube completely by coming up with the "solid state" system and the transistor—a long way from Marconi's invention. And I am sure that this admirable and inspired technical creation will itself be superseded by something better one day—perhaps tomorrow.

Where will the new atomic energy ideas come from? From established scientists? One would hope so—but that is unlikely. From some unknown but inspired amateur, some individual genius whose concepts will at first seem oddball to the professionals? Quite possibly. Perhaps the best ideas will come from a totally different area—from the industrial chemical laboratories, say, as was once suggested by the great industrial chemist Dr. Charles Thomas of Monsanto. Or they may emerge as a radical spin-off from the creative thinking of scientists in the fast-moving life sciences. Who knows where the new nuclear ideas may come from, stimulated by need, by youthful genius, by opportunity and encouragement.

The greatly respected dean of atomic energy research, Dr. Alvin Weinberg of Oak Ridge, has written recently, "I hope that those who believe that nuclear energy is too important to lightly cast aside will do more than simply restate their faith in the [prevailing] technology. They must come up with positive, new, and convincing initiatives that prove to the public that the lessons of Three Mile Island have been learned."

One of these lessons, in my view, is that a major new research and development enterprise should be begun at once, designed not merely to remove the "bugs" uncovered by Three Mile Island but to produce an entirely different technology. A new start.

In 1963, when the atomic energy industry was only a few years old, I appealed to the scientific and engineering leadership to reassess the headlong plunge into a greatly increased program of atomic reactors. I was by no means the only one who believed that the pace set by the industry and government was too precipitate to assure the long-term future of the technology as a principal source of energy for the country. What was lacking then—and remains lacking today—was a sense of prudence about *how rapidly* it was wise—or safe—to proceed toward that "new world" vision of this new source of energy.

As we enter the 1980s, atomic energy is a reality that demands to be dealt with responsibly, an essential element in our economic life and in the thus far chiefly verbal (and as yet ineffective) efforts to free this nation from dependence on foreign oil.

The failures of the present system of nuclear energy need not mark the end of the road. There is still time to change, still time to recapture the perspective and balance and excitement of new ideas that were lost along the way. We cannot afford to perpetuate the errors that we have already made; nor can we afford to abandon the entire effort as a bad job. We need to find a new way in atomic power, and I am certain that we will, in time.

How much time, no one can say. But time is needed, and patience, and perseverance, and substantial funding. It is still early; we are too near the beginning of the atomic age to make final judgments. The "new world" is yet to come.

# 13

# ATOMS FOR WAR:
# "ABSOLUTE NUCLEAR NONSENSE"

I have expressed views about the hazards to human health and safety of the present method of producing electricity from the nucleus of the atom. Yet a far deeper hazard—one that threatens the very survival of civilization as we know it—is rarely spoken of in the context of the present lively debate over peaceful atomic energy. Yet it is the very same science and technology that produces atomic electricity that has created the most terrible instruments of destruction the world has ever known, the thousands of atomic missiles that now menace the future.

This duality—the peaceful and unlimited energy so sorely needed the world over, and the capacity to produce by the same process the unlimited means of death and destruction—is the greatest paradox in history; it presents the world's political and religious leaders and institutions with the severest test of all time.

From the very beginning—in 1945—of the period in which, with many others, I have been wrestling with the atom, I expressed the conviction that this duality may well be the clue to a resolution of the tragic impasse; in the hopeful words of Shakespeare (from *Henry IV, Part I*), "Out of this nettle, danger, we pluck this flower, safety." It was this same duality that in February 1946, at the outset of the atomic age, underlay the Acheson-Lilienthal Plan, later the American plan for international peaceful atomic energy development and the elimination of national rivalry in atomic weaponry.*

In 1946, as chairman of a specially constituted Board of Consultants to the President and the Department of State, I and my board colleagues were called upon to design and propose a plan to further peaceful energy from the atom, worldwide, in such a way that we could move toward eliminating national rivalries in atomic weapons through an International Atomic Energy Authority.

The heart of the proposal would make the prospective benefits from the development of the peaceful atom an incentive and a means to the control and elimination of atomic arms.

Is it now too late to renew such efforts, even though today (March 1980) the world is closer than ever to an atomic arms confrontation between the United States and the Soviet Union?

I can make no attempt here, in this consideration of peaceful atomic energy, to explore thoroughly the present-day prospects of an up-to-date version of that earlier American initiative. Nevertheless, such a disclaimer is not entirely realistic. Atomic weapons are integral to the revolution of ideas that followed upon the atomic discoveries, and the two-sided nature of these discoveries is inherent in any consideration of atomic energy in the world today.

I have proposed that we make a new start toward a safer peaceful atom, using a technology that will not, as the present technology does, produce bomb material in the process of creating the peaceful atom.

I now propose that, without delay, we also consider a far more

---

*The complete text of the plan is contained in a State Department document, Publication 2498, 1946, substantially the same proposal that was presented as the American proposal to the United Nations on June 15, 1946, by the American delegation to the UN headed by Mr. Bernard Baruch.

important new start: to renew those earliest efforts to control and eliminate nuclear weapons, the one form of weapon that most endangers the world's future. The alternative is unthinkable: to resign ourselves to the soothing syrup of treaties that at best only stabilize desperately unstable confrontation.

The fact that peaceful atomic energy plants now exist throughout the world, that these plants do in fact produce plutonium suitable for use in atomic weapons, has multiplied and intensified the problem of atomic weapons controls that we faced when the United States alone was capable of producing atomic weapons. For some, the spread ("proliferation") of atomic weapons, far more than problems of plant safety or nuclear waste disposal, signals the death warrant for the once bright hope of 1947.

I do not agree with this doomsday view.

In some respects the picture today actually appears more hopeful than at almost any earlier time. It is a hopefulness that grows out of the desperate fact that all peoples, the Russian people included, are enmeshed in an identical circumstance that will bring disaster on all of us if it is not resolved affirmatively. It is a hopefulness reinforced by the pragmatic view that mankind has never before deliberately chosen a path that plainly and demonstrably leads to common disaster. I believe that the certainty of common disaster can produce a common accommodation.

Atomic weapons are the product of a great discovery, a discovery that has produced feasible, tangible, and greatly needed additions to the available and prospective energy supply of the world. The universal need for ever more energy to improve the lot of human beings is a major plus in the efforts to resolve the frantic atomic arms race. Turning people's minds to a consideration of atomic discoveries in their fruitful aspect could lead to a new effort that could produce, initially, a moratorium on the further development of atomic arms and, from such a pause, a movement away from the senseless, futile building up of stockpiles of nuclear weapons and the missiles that are the means of delivering them.

This is not merely wishful thinking. In 1947, to the surprise of most of the world, there was a broad-based affirmative reaction in this country and in every other country of the world, the Soviet Union excepted, to the proposals of the American plan of that time for dealing with atomic weaponry. The essence of the proposal was its affirmative aspect, the beneficial side of an international effort to develop an entirely new source of energy. The time for the revival of that basic idea with its broad appeal to the positive side of human nature may well not be here; but it may be around almost any corner, for the utter futility and the blood-chilling consequences of A-bomb rattling and the increasing risk of a terminal war are growing upon us.

If a safe *new* method of producing electricity from the atom were in the offing, a method that would not produce bomb material as the present method does, that prospect might provide the necessary leverage to bring about atomic weapons control.

No leader has stated more explicitly the urgency and immediacy of the risk of a continued multiplication of atomic bombs than Pope John Paul II on October 2, 1979, during his visit to the United States. The pope's utterances underscore my own reason for hope, namely, the realistic recognition that the ultimate danger in the immediate future of the world is embodied in the atomic arms race.

The pope spoke before the General Assembly of the United Nations:

We are troubled also by reports of the development of weaponry exceeding in quality and size the means of war and destruction ever known before. . . . The continual preparations for war demonstrated by the production of ever more numerous, powerful, and sophisticated weapons in various countries show that there is a desire to be ready for war, and being ready means *being able to start it;* it also means taking the risk that *sometime, somewhere, somehow, someone can set in motion the terrible mechanism of general destruction.* [Emphasis added.]

The Soviet leadership knows, and Americans have been told, the number and the capability of the atomic weapons that each nation possesses. These numbers are so overwhelming—they are in the tens of thousands—that to add more thousands to them could add nothing to anyone's security. Worse, the mountainous costs invite financial disaster for American and Soviet people alike, and the presence of the arsenals heightens the pressures to make use of them. In an hour's time much of the world, the accumulation of centuries, could be laid waste.

A moratorium might not yield immediate results, but the prospect of discussions among Soviets, Americans, and other peoples holds the possibility of our regaining the sense of sanity that the chilling and dehumanized poker game of the intellectuals of the arms controls business has almost destroyed.

I have alluded more than once to the negative effects of an excessive reliance on "experts" in the development of a safe and reliable source of peaceful atomic energy, to the exclusion of the ordinary citizen's views and insights. For a whole generation I have witnessed the rise of still another cult of experts, the arms control analysts. It is no more likely that they would question their own wisdom about atomic control than atomic energy experts would accept criticism of their views on nuclear plant safety. Both kinds of experts have failed us and failed the world.

The atomic control "experts" have developed esoteric theories of arms control that employ a vocabulary and a manner of thinking that for most people is incomprehensible jargon. What the average citizen knows well is that the numbers and the complexity of atomic weapons have increased; gone altogether is *any* sense of the security that atomic scientists, weaponeers, and analysts have promised, as they promoted one more round of new versions of atomic weapons, the "ultimate" weapon. Americans and Russians alike have been bilked; we and they have less security today than ever.

It is futile to ask the coterie of analysts, scientists, and arms control experts who dominate American and Russian military and political thinking on atomic warfare to admit their past failure, to admit the bankruptcy of their proposals to legitimize, by treaty or other means, even more advanced atomic weapons systems, to admit that the various doctrines of "mutual deterrence," "substantial equivalence," etc., etc., they have been feeding us for years no longer have credibility.

Are atomic weapons really designed for warfare? Are they in fact a part of any kind of military thinking? I remember how impressed I was thirty years ago, in a closed meeting with the American Joint Chiefs of Staff, to hear General Omar Bradley, a true soldier, declare that atomic weapons are not useful as military weapons at all. And, a short time before his tragic death, one of the world's leading military figures, Lord Mountbatten of Burma, said much the same thing: "As a military man who has given half a century of active service, I say that *the nuclear arms race has no military purpose. Wars cannot be fought with nuclear weapons.* Their existence only adds to our perils because of the illusions which they generate. . . . This is absolute nuclear nonsense." [Emphasis added.]

# 14

# A LOOK AHEAD

There's an old story about a traveler who stops his car on a country road to ask a farmer, "Can you tell me how to get to Smithville?" The farmer thinks for a minute, then says, "Well, if I wanted to get to Smithville, I wouldn't start from here."

If we want to reach our nuclear energy destination, we will have to realize that we cannot get there from where we are.

Nuclear energy is by no means finished; it remains one of the great hopes of mankind, and in due course it will play a major role, perhaps the decisive role in providing the energy that the world needs so badly. But that goal will not be reached on the road we are now traveling. We need to back away from our present nuclear state in order to find a better way, a route less hazardous to human health and to the peace of the world and its very survival.

The decade of the 1980s will be crucial not only for the future of atomic power but for the broader issues that have been summed up and illuminated by the dramatic worldwide nuclear controversy. Never before have we been so gravely menaced by what our science and technology have created—and what the powers and principalities of the earth have so signally failed to put to use and control in a decent and humane way.

The citizen protest against atomic energy plants here and abroad was not raised against nuclear hazards alone. Nor would it be satisfied if all nuclear plants were to be closed tomorrow. To a large extent it has been a protest against the misuse of science, the misdirection of enormous forces that human ingenuity has brought into being. It is a protest against the abuses of industrial technology that poison the land instead of nurturing it, that sour the air and foul the water, that devour marsh and woodland and make hazardous to health and peace of mind the cities and factories in which people live and work. It is a protest against governments—all governments—that spend billions in an endless, insane atomic arms race that consumes the cream of the world's resources and much of its brightest human talent. It is a protest against the uses of science and technology that are antihuman and antilife.

This is not to say that today's protesting citizen groups do not include among them individuals who are arrogant, ignorant, and self-seeking. Many of the demands and arguments that they put forward are plainly exaggerated, even hypocritical.

Nevertheless there is a strong, genuine, largely spontaneous surge of feeling that animates these protests, a feeling that expresses, in crude ways at times, the anxieties of a great part of the public. I sense that people in general know that something has gone wrong not just with nuclear power but with the industrial and technological forces that produced it in its present hazardous form. These misgivings are part of a larger sense of worry and unease about our society itself. Ordinary citizens find their well-being threatened, their serenity shattered; they have doubts about the food they eat, the air they breathe. They are aware of a growing frustration, a feeling that it is time to call a halt. Where better to call a halt than with the most mysterious and dramatically menacing of all forces, the atom?

Atomic energy was heralded at the outset as the brightest hope for a future of abundant power. Instead it has added hazards to a world that already had too many.

It is painful to reflect on the road we have traversed in this first generation of the atomic age. In the beginning we had no bombs and no safe and effective way to turn nuclear energy into electricity. Now there are more than plenty of A-bombs—and still a less than safe and effective way to turn nuclear energy into electricity.

The fact that we do not have the right answer now does not mean that we will not have it one day. But we certainly will not have it if we tell ourselves that the present faulty method only needs to be tinkered with, to be shored up, to have better-trained operators and still more safety backup systems plugged into it.

*We have to make a new start* in order to find a better, cleaner, safer way of producing electric energy from the atom.

Now, energy for what and for whom?

No American energy policy makes sense unless it takes into account the hopes and needs of *all* peoples. Our international political position has been damaged by the assertions of our own introspective energy "experts" who have branded us heedless gobblers of resources, particularly energy resources, when it is these very energy resources that have enabled us to aid the helpless the world over.

"Conservation" is advanced passionately as the answer to America's energy problem. If conservation means minimizing obvious waste and increasing efficiency, how can anyone possibly oppose such a self-evident proposition? But if conservation is intended as the principal means of achieving American energy self-sufficiency—as some assert—isn't this nothing more than a kind of isolationism in a particularly heartless form, an elitist disguise to mask putting a limit on total energy production, thereby slowing economic growth for those who need growth the most? In urging a conservation policy of "making do with less and less," we need to remember that most of the world's people have little or no energy to "conserve."

We grumble when there is a temporary shortage of gasoline for our cars, when a brownout briefly stalls elevators and melts the ice cubes in the freezer. But the overwhelming majority of human beings have no gasoline, no cars, no elevators, no refrigerators. Their countries have never had much energy, they do not have much today, and they will not have much more in the foreseeable future.

It requires little imagination to picture what goes on in the mind of an individual from one of these countries who visits the United States and overhears our talk of an energy crisis while observing our cities lighted up like giant jukeboxes, our highways choked with cars and trucks, and all the other evidence of our powerhouse civilization.

Our future energy policy may be a technical success, but it will be a moral and political failure if it aims only at filling our gas tanks and keeping our thermostat settings high. Such a limited, nationalistic energy policy will inevitably pit us, the richest of the rich, against the less developed nations in the greedy business of grabbing everything we can for ourselves. And because there just isn't enough to go around, and there probably never will be enough, our ill-tempered reactions to minor energy shortages will worsen. We then run the risk of yielding to the recurrent proposals to ready the Marines for the seizure of the oil fields of the Persian Gulf states.

The only escape from this sinister and destructive meanness is to frame policies that recognize our ethical responsibility toward the other peoples on this planet as fellow members of a world community. This task cannot be dismissed as empty idealism; it is the only practical policy for the long run. Unless we make a major contribution toward easing the *world's* energy shortage —instead of merely satisfying our own needs—we may be creating for our children's children a life of constant crisis and chronic insecurity.

And our own view of ourselves—our national soul—will suffer.

During my lifetime I have witnessed with pride a series of creative and productive advances in America, each greater than the one before. There is still poverty in this country, but the general standard of living has risen to a level undreamed of when I was a boy. I remember the stores in Valparaiso, Indiana, around 1910; the customer back then had a modest range of things to buy and not much choice. Today a shopper at one of our gigantic malls finds an abundance and variety of goods that would have struck dumb with wonder the Main Street shopkeepers of my boyhood.

The American dream of plenty has come true.

Yet there is more to that dream than plenty. There must be, for otherwise we are no more than what our detractors say we are, an overfed, self-indulgent nation of consumers. I do not believe this is true. Every day we see strong evidence that we are as capable now as we have been in the past of seeing beyond our own needs, of looking with understanding and acting with fairness toward other peoples.

This I do believe: our energy policy will be a test of our moral worth and our place in history. It will be a sign and a testament; it will show the true face of America to the world; it will tell us, too, what kind of people we really are.

With the other nations of the world we share a common problem, the necessity for finding the energy that today and in the years ahead we all need so urgently. In the past twenty years we have developed a technical system for converting nuclear fission to electricity, and we have made this particular technical system—the light water reactor—dominant in the international energy market. Now, belatedly, we recognize that our nuclear technology is not really so advanced. It isn't dependable enough, it isn't safe enough. And we have the responsibility to ourselves and to other peoples and nations to begin at once to create a better answer, no matter how long it takes. When we have it—and we *will* have it some day—we will have satisfied a historic obligation as well as a desperate necessity. It would be most appropriate if the nation that made and used the first atomic bomb should be the nation that creates the first safe atomic power.

Then we will have one more American product well worth marketing, a product that we can stand behind as proudly as any of my hometown storekeepers when they offered an honest and reliable piece of goods to their customers.

True, in competent hands we are able to make the present system work reasonably well. We have engineers who are capable of devising extra backup safety measures, and in the TVA we have an example of top-notch management able to train teams of superoperators to handle the heavy demands of running what the engineers put together. Through skill and ingenuity we can live with and accommodate ourselves to the defects of our present technology, as most Western European nations and Japan believe they are able to do. But what about the poorer, less advanced nations, those who don't have the technical skills and trained human resources that we have? From my professional work in faraway countries, I know that neither now nor in the early future will the underdeveloped countries be able to juggle the complicated risks that we can barely manage—yet their need for energy is many times greater than ours.

In recent years I have asked myself often, do we Americans have the *moral* right to promote and sell a complicated, immature, and fundamentally unsafe nuclear system to the rest of the world, particularly to societies whose technology is far less developed than our own?

I am convinced that we do not have that right.

We *do* have the obligation to develop a nuclear energy system that is far safer, much less complex, for them and for ourselves.

International cooperation is mutual exchange, in energy or anything else, and we have much to learn from others. Brazil, for example, is pioneering in a big way in producing alcohol from sugar plants to add to motor fuel. Italy is developing advanced means for the utilization of solar energy and of hot air geysers, or thermal springs. Even in the nuclear field some other nation may be the first to find a clean way of producing nuclear power. After all, it was European scientists domiciled in our country who were the pioneers in applied nuclear physics.

In the present period of increasing fears about nuclear hazards and the adequacy of oil supplies, energy has become something of an international monster, a source of hatred and suspicion even among friends. Surely this is not an immutable situation. International cooperation on common problems has not accumulated an overwhelmingly fat dossier of successes, but past examples there have been, and present ones exist as well, both at regional and world levels. The European Common Market is an important case; so are some notable specialized United Nations agencies, among them the World Health Organization, the Food and Agriculture Organization, and the World Bank. In the scope of the present writing I can do no more than to add my voice to those who recognize the fundamental fact: the energy problem is a *world* problem, and it cannot over time be confronted effectively by anything less than a *world* point of view.

The energy problem could become a source of international cooperation and hope. It will not, however, as long as the highest aim of our energy policy remains one of national self-sufficiency. Certainly we do not want to live by the sufferance of some international cartel; no people should have to be thus subservient. But our self-sufficiency in energy should not be crassly achieved at the expense of less powerful nations. How can we guarantee that this will not be the case unless we participate in common action to assure fair shares of energy for all?

The best way to achieve plenty of energy for all is through the creation of more energy. Safe energy.

Research and development of "clean" nuclear energy could be a starting point for international cooperation in that area and in other parts of the world energy picture.

Such cooperation might lead—might be made to lead—to some easing of the threat of nuclear weapons. An American-sponsored moratorium on the production of more bombs could conceivably provide the breathing space for a new start, perhaps a revival in some altered form of the American Plan for International Control of Atomic Energy proposed by this country more than thirty years ago. That plan was based on the all-out world-wide development of the *peaceful* atom, using the world's great need for more energy as leverage and incentive for the control and elimination of the *dangerous* atom, the bomb.

# EPILOGUE

I began these reflections on energy by musing on the changes that have occurred since my boyhood in a small midwestern town—a transformation without precedent, an historic transformation sparked by an era of energy.

Whether all these changes have been for the good may be doubted. What is beyond doubt is that for the first time in history, millions of ordinary people can hope to have, through energy, a greater share of the things of this world. And not alone the physical comforts—food, shelter, the easing of drudgery—but the intangible treasures of life as well, some security, a little leisure and privacy, some respite from anxiety . . . something of the serenity I have found on the remote hilltop where I have been writing these words during the past summer months.

What I have reflected upon and written about is not merely a new source of electrical energy, nor energy as an economic statistic. My theme has been our contemporary equivalent of the greatest of all moral and cultural concerns—fairness among men and the endless search for a pathway to peace.

# INDEX